The Role of Plant Pathology in Food Safety and Food Security

Plant Pathology in the 21st Century

For other titles published in this series, go to
www.springer.com/series/8169

R.N. Strange • Maria Lodovica Gullino
Editors

The Role of Plant Pathology in Food Safety and Food Security

Springer

Editors
R.N. Strange
School of Biological and Chemical Sciences
Birkberk College
University of London
UK
r.strange@sbc.bbk.ac.uk

Maria Lodovica Gullino
Universitá degli Studi di
Torino, Italy
marialodovica.gullino@unito.it

Low yield of a banana plant derived from a sucker (lower picture) and higher yield of banana plant grown from tissue culture (upper picture): see article by James Onsando and Florence Wambugu (page 80).

ISBN 978-94-007-3108-0 e-ISBN 978-1-4020-8932-9
DOI 10.1007/978-1-4020-8932-9
Springer Dordrecht Heidelberg London New York

Cover illustration: Photograph courtesy of Dr. James Onsando

Printed on acid-free paper

Springer is part of Springer Science+Business Media (www.springer.com)

Foreword

This collection of papers represents some of those given at the International Congress for Plant Pathology held in Turin in 2008 in the session with the title "The Role of Plant Pathology in Food Safety and Food Security". Although food safety in terms of "Is this food safe to eat?" did not receive much direct attention it is, nevertheless, an important topic. A crop may not be safe to eat because of its inherent qualities. Cassava, for example, is cyanogenic, and must be carefully prepared if toxicosis is to be avoided. Other crops may be safe to eat providing they are not infected or infested by microorganisms. Mycotoxins are notorious examples of compounds which may contaminate a crop either pre- or post-harvest owing to the growth of fungi. Two papers in this book deal with toxins, one by Barbara Howlett and co-workers and the other by Robert Proctor and co-workers. In the first of these, the role of sirodesmin PL, a compound produced by *Leptosphaeria maculans*, causal agent of blackleg disease of oilseed rape (*Brassica napus*), is discussed. The authors conclude that the toxin plays a role in virulence of the fungus and may also be beneficial in protecting the pathogen from other competing micro-organisms but there seem to be no reports of its mammalian toxicity. In the second paper, attention is drawn to the many plant diseases caused by the fungus, *Fusarium proliferatum*, and the fact that it produces a wide range of biologically active metabolites, including the fumonisins. These compounds, in a related fungus, *Gibberella fujikuroi*, play a role in virulence but have also been identified as carcinogens.

The remaining chapters of the book deal with the security of our crops and how this may be improved. In the first paper, Ziegler and Savary document the importance of rice, particularly for the poor, and point out that the recent price spike has been catastrophic for them as, even before the increase, many were spending at least half their budget on food. Regarding the toll taken by disease, this is difficult to estimate but under current agronomic conditions grain yield reductions are thought to be about 15%.

The paper by Emmanuel Moses was concerned with the improved yields of cassava in cultivars that are resistant to African Cassava Mosaic Disease and, with the help of a grant from the International Society for Plant Pathology, the education of farmers in the recognition of diseases and their avoidance by using healthy material for vegetative propagation. Yield increases have been spectacular, double those of the yields previously obtained in some instances.

Several chapters deal with quarantine measures. Increases in trade in plants and their products carry increased risks of importing dangerous pathogens. The paper by van der Graaff and Khoury goes further in that it encompasses "the policy and regulatory frameworks (including instruments and activities) for analyzing and managing relevant risks to human, animal and plant life and health, and associated risks to the environment" as well as "food safety, zoonoses, the introduction of animal and plant diseases and pests, the introduction and release of living modified organisms (LMOs) and their products (e.g. genetically modified organisms or GMOs), and the introduction and management of invasive alien species".

Scott and Strange recount the involvement of the International Society for Plant Pathology's concern for food security from the 1998 Congress to the present. Following an address by the Nobel Laureate, Norman Borlaug, at that Congress, a Taskforce for Food Security was formed and a meeting took place in Bangkok the following year. Other significant developments include the funding of Emmanuel Moses's project (mentioned above) and the launch of the new journal, "Food Security: the Science, Sociology and Economics of Food Production and Access to Food," in association with the publisher, Springer.

With the increase in global travel it is inevitable that there has been movement of plant pathogens across evolutionary and geophysical boundaries as described by Evans and Waller. Their paper includes some measures that could be taken to improve phytosanitary control. However, it is conceivable that malicious individuals, disaffected groups of people or the political elite of a country might contemplate using plant pathogens in agro-terrorism, as pointed out by Stack and co-workers. Reassuringly, they conclude that it is difficult to establish a pathogen in a crop at a distance where it will multiply to give a serious plant epidemic.

Preventing the introduction and spread of plants and plant products as well as controlling them is the purpose of the International Plant Protection Convention (IPPC) as described by Roberts. Encouragingly, there are now 170 contracting parties to the IPPC. Similar objectives are also espoused by the European and Mediterranean Plant Protection Organization (EPPO) as described by Petter and co-workers. In addition EPPO has been developing Pest Risk Analysis (PRA) for specific plant pathogens.

Onsando and Wambugu point out the benefits that have resulted from exploiting biotechnology for plant disease control with several examples ranging from the establishment of healthy tissue culture (see cover of this book) to genetic modification. The latter still excites considerable opposition in Europe where those opposed would, it seems, prefer conventional plant breeding in which many flanking genes besides the one of interest are transferred. As Jonathan Gressel points out in his book, "Genetic Glass Ceilings" (published 2008), breeding is like getting a spouse with a whole village whereas genetic engineering is like getting a spouse without in-laws! Finally, Hohn, and Schachermayr consider genetic modification (GM) as a new tool in the resistance toolbox and present some exciting results in which plants have been transformed with genes conferring resistance to aphids and, through gene silencing, resistance to viruses.

Richard Strange

Contents

The Role of Plant Pathology in Food Safety and Food Security

Plant Diseases and the World's Dependence on Rice 3
R.S. Zeigler and S. Savary

Development of Appropriate Strategies to Control Cassava Diseases in Ghana 11
E. Moses

Biosecurity in the Movement of Commodities as a Component of Global Food Security 25
N.A. van der Graaff and W. Khoury

Global Food Security

ISPP and the Challenge of Food Security 43
Peter Scott and Richard Strange

Globalisation and the Threat to Biosecurity 53
Harry C. Evans and James M. Waller

Genetic Modification (GM) as a New Tool in the Resistance Toolbox 73
Thomas Hohn and Gabriele Schachermayr

The Role of Plant Pathology and Biotechnology in Food Security in Africa 77
James M. Onsando and Florence Wambugu

Mycotoxins

The Secondary Metabolite Toxin, Sirodesmin PL, and Its Role in Virulence of the Blackleg Fungus 89
Barbara J. Howlett, Ellen M. Fox, Anton J. Cozijnsen, Angela P. Van de Wouw and Candace E. Elliott

Biological and Chemical Complexity of *Fusarium proliferatum* 97
Robert H. Proctor, Anne E. Desjardins and Antonio Moretti

Biosecurity and Quarantine

Bioterrorism: A Threat to Plant Biosecurity? 115
J.P. Stack, F. Suffert and M.L. Gullino

The Revised International Plant Protection Convention – a New Context for Plant Quarantine 133
William Roberts

Pest Risk Analysis as Applied to Plant Pathogens 137
Françoise Petter, Sarah Brunel and Muriel Suffert

Index 151

Part I
The Role of Plant Pathology in Food Safety and Food Security

Chapter 1
Plant Diseases and the World's Dependence on Rice

R.S. Zeigler and S. Savary

1.1 Introduction: Unprecedented Agricultural Changes and Challenges, and the Importance of Rice Diseases

This brief review deals with rice, the first food crop of humankind, and with rice diseases. It also deals with our present and future need to manage rice diseases in a sustainable way. Therefore, we also touch upon yield-reducing factors other than diseases within the tremendous diversity of contexts in which farmers grow rice. It is this agricultural, social, ecological, and economic diversity that science and research policies must address in an unprecedented global context of rapid changes and challenges.

September 2007 to July 2008 saw an extraordinary increase in global rice prices, with a peak price surpassing US\$750 t^{-1} in February 2008. The main causes for this spike include (S. Pandey, IRRI, personal communication): (1) the inelasticity of rice trade worldwide, compared to other cereals (whose prices also increased very strongly during that period), (2) very low national rice stocks associated with plateaus in yield growth and cultivated rice area expansion rates, and (3) competition among cereals to meet global and national demands for food, feed, and energy (biofuels). There is debate, of course, about the specific contributions of various factors to this price spike, but the fact remains that, ultimately, consumers had to pay more for their food. Although this price increase may have been relatively benign to wealthy consumers, it was not so for the world's poor: as food accounts for at least half of their daily budget (Zeigler 2001), a twofold increase in price is simply unbearable.

R.S. Zeigler
International Rice Research Institute, Los Baños, Laguna, Philippines; International Rice Research Institute, DAPO Box 7777, Metro Manila, Philippines
e-mail: r.zeigler@cgiar.org

S. Savary (✉)
International Rice Research Institute, Los Baños, Laguna, Philippines; Plant Pathology, IRRI, DAPO Box 7777, Metro Manila, Philippines
e-mail: s.savary@cgiar.org

R.N. Strange and M.L. Gullino (eds.), *The Role of Plant Pathology in Food Safety and Food Security,* Plant Pathology in the 21st Century 3,
DOI 10.1007/978-1-4020-8932-9_1, © Springer Science+Business Media B.V. 2010

The vast majority of the poor worldwide – nearly 70% of them (Zeigler and Barclay 2008; R. Hijmans, IRRI, personal communication) – live in Asia, and depend on a predominantly rice-based diet (Smil 2000). Another large fraction of the poor worldwide lives in Africa, where rice is also a key staple. Although there is debate on the mechanisms that led to the 2008 spike in cereal prices, especially for rice, there is no debate on the fact that the spike built upon a steady price increase that started in 2001 (Zeigler and Barclay 2008; S. Pandey, IRRI, personal communication). There is no debate either on the fact that this was partly because the rate of global rice production has been slightly lower than 1% during the past decade, whereas it should surpass 1.5% to meet global food needs now and until at least 2020 (Greenland 1997; Daily et al 1998). Herein lies the true challenge we face: rice productivity must increase, not only because of the arithmetic needs of a growing global population (Dyson 1999) but also because more elasticity in rice production–consumption must prevent recurrence of future crises. Any factor that limits attainable yields, or that reduces actual yields, will hamper progress toward achieving both this necessary rate of increase in yield and this higher elasticity. Plant diseases are key yield-reducers. Improving rice crop health in a sustainable manner therefore can, and must, contribute to achieving global food sufficiency and reducing poverty.

1.2 The Importance of Rice Diseases as Yield-Reducers

The current importance of rice diseases and of rice yield-reducing factors in general is fairly well documented in the lowlands of Asia, which account for the bulk of the world's rice production. Considering a current average attainable yield of 5.5 t ha^{-1}, a 10-year study conducted by IRRI and its partners (Savary et al. 2000a, b) in several hundred fields across Southeast and South Asia indicates that yield-reducing factors cause 37% losses. Among these, three diseases are responsible for losses of 5% or more each: sheath blight, brown spot, and blast. This is, on average, much more than insect-related losses (0.3%, 0.1%, and 2.3% caused by leaf-feeding insects, stem borer 'deadhearts', and stem borer 'whiteheads', respectively), and this is, as expected, much less than weed infestation (approximately 20% losses). Simulation modeling (Willocquet et al. 2004) indicates that, overall, yield-reducing factors cause an average yearly yield attrition ranging from 120 to 200 × 10^6 t of grain over the 87 × 10^6 ha of lowland rice of tropical Asia. Disaggregating these figures by diseases is hard, for two reasons. One is that yield losses to diseases, and pests in general, have long been known to interact (Padwick 1956) – they are less than additive in most cases; thus, the individual contribution of a particular disease to an overall yield loss mean is often attached to the wrong assumption that removal of the injury would translate into a corresponding gain in yield. A second reason is that injuries vary strongly across production contexts: the means cited above cannot reflect the wide variation in injury intensities. Arithmetic means across large data sets do not do justice to extreme events, such as devastating virus epidemics, which can be captured only incompletely in a survey of such size. Rice production globally, for example, is constantly exposed to virus epidemics, some of which, such as rice tungro in Asia, have the potential to destroy a crop entirely (Savary et al. 2000b). Such catastrophic events overlay their yield-reducing effects on chronic, near-omnipresent diseases such as rice sheath blight. Simulation and experimental work indicate that rice sheath blight alone is responsible for yield losses of 6% across the irrigated and rainfed lowland rice systems of tropical Asia (Savary et al. 2000b; Willocquet et al. 2004). It is nevertheless a fair assumption that rice diseases cause 15% rice grain yield reductions under the current contexts of both crop production and disease management.

Simulation modeling enables us to assess the performance of disease management tools and speak of yield gains instead of yield losses. Although currently deployed management tools for diseases are quite inefficient for rice sheath blight (45–60% control efficiency at best), they are very good (95%) to excellent (nearly 100%) for blast and bacterial leaf blight, respectively. Such very high management efficiencies correspond to diseases against which host–plant resistance research, breeding, and deployment have had great success. Simulation modeling also enables us to contemplate the losses diseases might reach if continuous research for new resistance genes and their deployment were to stop. Today, bacterial leaf blight and blast cause yield losses of around 0.1%. Were host–plant resistance not deployed in currently cultivated varieties, losses in excess of 5% for each disease would be expected, with devastating consequences.

1.3 Research to Prevent Losses to Specific Diseases

New strategies are being developed to identify new resistance genes, with a two-pronged approach. One component is based on the systematic exploitation of existing genetic variation in rice germplasm, which enables us to dissect characters to identify candidate genes that are expressed in backcrosses and near-isogenic lines; another is based on developing a large collection of mutants, which are then screened for gain or loss of resistance. Combined, the two approaches provide convergent evidence for host–plant resistance genes and the development of a candidate gene pool. This, in turn, opens the way toward association genetics, while advanced backcross lines enable us to evaluate consensus candidate genes (Leung et al. 2001). This approach is now used for both blast and bacterial leaf blight. For instance, for bacterial leaf blight, work started in the late 1970s with the identification of resistance gene donors and was continued in inheritance studies by gene identification and the development of near-isogenic lines in the 1990s (Mew and Vera Cruz 2001), in combination with pathogen population analysis (Leach et al. 1992), understanding of pathogen genes underpinning aggressiveness (Bai et al. 2000; Vera Cruz et al. 2000), and the identification of markers of resistance genes (Leach et al. 2001). These studies led to the development and release of new resistant varieties in the 2000s. This example shows how basic research can translate into a better understanding of processes and lead to major, yet insufficiently recognized, impacts for farmers: were it not for the simulation modeling work, no one could tell that this research led to yield gains of 5–15% across Asia. The example also gives a measure of the timeline attached to any basic research required to achieve impact: assuming that the full deployment of a new rice variety takes a minimum of 10 years, 40 years of continuous support were required for the fundamental research on the diversity of the causal agent of bacterial blight, *Xanthomonas oryzae* pv. *oryzae,* and host resistance genes to be developed into practical applications. Actually, a delay of about 40 years between the inception of research and its practical application is within the range of estimates given by economists specializing in the global benefit of agricultural research (Pardey et al. 2006; Pardey and Wood 2008).

Fundamental science does not reside only at the molecular level. Basic understanding on the spread of disease epidemics during the late 1960s (Gregory 1968), on the epidemiological implications of genetic uniformity (Browning 1974), and on the dynamics of epidemics (Zadoks 1971) in the 1970s led to new ideas on how to engineer host genetic landscapes that are more resilient to plant pathogens (Wolfe 1985). The idea was then applied to the management of blast, arguably the most important, and the most genetically shifty, rice disease. It was implemented in Yunnan, China, initially as a modest collaboration between Chinese research institutions and IRRI. Several trials were attempted in the early 1990s, unsuccessfully. The need to simultaneously grow rice as food with an acceptable yield and to grow rice as a cash crop too – as a delicacy that is associated with ancient customs – led to a practical solution (Zhu et al. 2000). Four to six rows of blast-resistant indica hybrids separating a single row of japonica glutinous rice not only controlled blast but also ensured acceptable food-yield and generated increased yield-income to farmers, who are among the poorest people in China (Revilla et al. 2001). The experiment started with a few plots, each a μ (0.15 ha) or less wide. Today, about 40% of Yunnan's cultivated rice area is using such a genetic association.

This second example again illustrates the necessary time lag, about 40 years, between basic fundamental research and impact. The practical impact of scientific advances cannot be guaranteed, but the two examples show that planning for impact can be, and actually is, inherently part of scientific advances. Plant pathology exists because of the perspective of such applications, and that is probably a major asset of the discipline.

1.4 Dealing with Crop Health Syndromes

Diseases do not occur in isolation, as we indicated in the early part of this brief review. In essence, rice in tropical Asia is exposed to three broad syndromes, that is, three main types of disease associations:

- Syndrome 1: Sheath blight and stem rot (along with plant hoppers, stem borer injury, and weed infestation)

- Syndrome 2: Bacterial leaf blight and some stem rot, sheath rot, sheath blight, brown spot, and leaf blast (along with defoliating insects, some stem borer injury, and very high weed infestation)
- Syndrome 3: Leaf blast, neck blast, brown spot, sheath rot, and some sheath blight (along with stem borer injury and some weed infestation)

These are only broad generic patterns, emerging from multi-year, multi-site characterization work (Savary et al. 2000b, 2006). Although these patterns may not exactly fit each of the several hundred farmers' fields surveyed, they give us a current overview. Two main findings of this characterization work are that (1) these syndromes are not site-specific: they can occur in locations several hundred micrometer apart, and (2) such crop health syndromes are very strongly dependent on patterns of cropping practices and, more generally, on production situations.

Although some farmers may be particular about a few specific diseases, such as blast or bacterial leaf blight, all farmers are concerned about crop health as a whole. A first challenge is therefore to design production contexts in which crop health syndromes would not compromise agricultural sustainability. Certainly, host–plant resistance is one component of success, but designing disease-risk-acceptable, sustainable rice production systems will also take a better understanding of the network of interactions among syndromes and production situations.

Because of shrinking agricultural water, declining rural labour, and increasing costs of fuel and fertilizer, agricultural change is taking place at an unprecedented pace in the world's rice-producing countries. This change in production situations also translates into changes in the relative importance rice diseases (and rice yield-reducing factors in general) may have (Savary et al. 2006). A second challenge is therefore to design new production situations so that their vulnerability to new crop health syndromes would be minimized while the consequences of climate change and globalization unfold. Here again, host–plant resistance will be an important ingredient, all the more so because the new crop health syndromes are very likely to include 'newcomers', or diseases that were until now considered secondary in importance. Sudden outbreaks of diseases, such as the rice yellowing syndrome in 2007–2008 in Southeast Asia (I.R. Choi, IRRI, personal communication), are also to be expected.

1.5 Concluding Comments

When crises such as the current one unfold, agricultural scientists in general and plant pathologists in particular, inevitably refer to the need for science to generate the required response. Experience shows, however, that impact cannot be guaranteed if not grounded in long-term efforts and the accumulation of knowledge that can be mobilized. Immediate needs should not compromise long-term investment. For example, the dramatic price spike of rice in February 2008 masked the upward pressure on prices that went unnoticed for half a decade. Balancing the need for rapid impact and the necessary scientific investment can be difficult, unless strategic programs are developed. Plant pathologists are actually accustomed to dealing with these sudden seemingly erratic outbreaks – while yield losses of some 37% continue unabated season after season. Although sudden outbreaks can immediately put at risk millions of livelihoods, systematic yield attrition causes low yield, and yield instability, which both contribute to poverty, especially in South Asia and Africa. Scientific investment is not only material or technical but also human. As this manuscript is being submitted, IRRI with its African partners and with WARDA (the Africa Rice Center) will organize a new training program for breeders and plant pathologists working in eastern and south-eastern Africa. We see the building of human capital and the sharing of knowledge as powerful devices for designing scientifically sound research programs with strong practical impact.

Acknowledgement We thank Bill Hardy, Communication and Publications Service, IRRI, for editing the manuscript.

References

Bai JF, Choi SH, Ponciano G, Leung H, Leach JE (2000) *Xanthomonas oryzae* pv. *oryzae* avirulence genes contribute differently and specifically to pathogen aggressiveness. Mol Plant Microbe Interact 13:1322–1329

Browning JA (1974) Relevance of knowledge about natural ecosystems to development of pest management systems for agro-ecosystems. Proc Am Phytopathol Soc 1:191–199

Daily G, Dasgupta P, Bolin B, Crosson P, du Guerny J, Ehrlich P, Folke C, Jansson AM, Jansson BO, Kautsky N, Kinzig A, Levin S, Mäler KG, Pinstrup-Andersen P, Siniscalco D, Walker B (1998) Food production, population growth, and the environment. Science 281:1291–1292

Dyson T (1999) World food trends and prospects to 2025. Proc Natl Acad Sci USA 96:5929–5936

Greenland DJ (1997) The sustainability of rice farming. CABI, Wallingford, UK, and International Rice Research Institute, Manila, Philippines

Gregory PH (1968) Interpreting plant disease dispersal gradients. Annu Rev Phytopathol 6:189–212

Leach JE, Vera Cruz CV, Bai J, Leung H (1992) Assessment of genetic diversity and population structure of *Xanthomonas campestris* pv. *oryzae* with a repetitive DNA element. Appl Environ Microbiol 58:2188–2195

Leach JE, Leung H, Wang GL (2001) Dissection of defense response pathways in rice. In: Rice biotechnology: improving yield, stress tolerance and grain quality. In: Goode, JA and Chadwick, D (eds) Novartis Foundation symposium 236, Wiley, New York, pp 190–204

Leung H, Wu C, Baraoidan M, Bordeos A, Ramos, M, Madamba S, Cabautan P, Vera Cruz C, Portugal A, Reyes G, Bruskiewich R, McLaren G, Laffitte R, Gregorio G, Bennett J, Brar D, Khush G, Schnable P, Wang G, Leach J (2001) Deletion mutants for functional genomics: progress in phenotyping, sequence assignment, and database development. In: Khush GS, Brar DS, Hardy B (eds) Rice genetics IV. Science Publishers, New Delhi, India, and International Rice Research Institute, Manila, Philippines, pp 239–251

Mew TW, Vera Cruz CM (2001) Bacterial blight of rice. In: Maloy OC, Murray TD (eds) Encyclopaedia of plant pathology. Wiley, New York, pp 71–74

Padwick GW (1956) Losses caused by plant diseases in the tropics. Commonwealth Mycological Institute, Kew, Surrey, Phytopathology Papers No. 1

Pardey PG, Wood S (2008) Linking plant diseases and their economic consequences in a global setting. Phytopathology 98(6):S188 (Abstr.)

Pardey PG, Alston JM, Piggott RR (eds) (2006) Agricultural R&D in the developing world: too little, too late? International Food Policy Research Institute, Washington, D.C., 398 pp

Revilla IM, Fan JX, Zhu YY, Li S, Mew TW, Hossain M (2001) Adoption of mixture planting for diversity in China: its impact on pest management and farmers' income. In: Mew TW, Borromeo E, Hardy B (eds) Exploiting biodiversity for sustainable pest management. International Rice Research Institute, Los Baños, Philippines, pp 217–236

Savary S, Willocquet L, Elazegui FA, Teng PS, Du PV, Zhu D, Tang Q, Lin X, Singh HM, Srivastava RK (2000a) Rice pest constraints in tropical Asia: characterization of injury profiles in relation to production situations. Plant Dis 84:341–356

Savary S, Willocquet L, Elazegui FA, Castilla NP, Teng PS (2000b) Rice pest constraints in tropical Asia: quantification of yield losses due to rice pests in a range of production situations. Plant Dis 84:357–369

Savary S, Teng PS, Willocquet L, Nutter FW Jr (2006) Quantification and modeling of crop losses: a review of purposes. Annu Rev Phytopathol 44:89–112

Smil V (2000) Feeding the world. A challenge for the twentieth century. The MIT Press, Cambridge, MA, 360 pp

Vera Cruz CM, Bai JF, Oña I, Leung H, Nelson RJ, Mew TW, Leach JE (2000) Predicting durability of a disease resistance gene based on an assessment of the fitness loss and epidemiological consequences of avirulence gene mutation. Proc Natl Acad Sci USA 97:13500–13505

Willocquet L, Elazegui FA, Castilla NP, Fernandez L, Fischer KS, Peng S, Teng PS, Srivastava RK, Singh HM, Zhu D, Savary S (2004) Research priorities for rice disease and pest management in tropical Asia: a simulation analysis of yield losses and management efficiencies. Phytopathology 94:672–682

Wolfe MS (1985) The current status and prospects of multiline cultivars and variety mixtures for disease resistance. Annu Rev Phytopathol 23:251–273

Zadoks JC (1971) Systems analysis and the dynamics of epidemics. Phytopathology 61:600–610

Zeigler RS (2001) Agricultural biotechnology: reducing poverty in developing countries. Plant Dis 85:568–579

Zeigler RS, Barclay A (2008) The relevance of rice. Rice 1:3–10

Zhu Y, Chen H, Fan J, Wang Y, Li Y, Chen J, Fan JX, Yang S, Hu L, Leung H, Mew TW, Teng PS, Wang Z, Mundt CC (2000) Genetic diversity and disease control in rice. Nature 406:718–722

Chapter 2
Development of Appropriate Strategies to Control Cassava Diseases in Ghana

E. Moses

2.1 Introduction

Cassava (*Manihot esculenta* Crantz) is the only member of the family Euphorbiaceae that is cultivated as a food crop (Fauquet and Fargette 1990). It is the third largest source of carbohydrate for human consumption in the world and the most important food crop in Africa. More than 80% of cassava produced in the world is consumed by human beings and it is the principal carbohydrate source for more than 500 million people in the tropical world (Lozano 1986; Fauquet and Fargette 1990). The cassava storage root contains 92% carbohydrates on a dry weight basis. Leaves of cassava contain 7% proteins on a fresh weight basis and are used as vegetables in Ghana and several other African countries.

Cassava is the most important food crop for the majority of the 22 million people who live in Ghana. The storage roots can be processed into several food forms including 'gari', tapioca and 'konkonte' that can be stored for 12 months or longer without loss of quality. In urban and rural Ghana, processed cassava food products contribute immensely to food security, especially in the lean season where other food crops may not be available or may be too expensive for the urban and rural poor. They are also popular in Nigeria and several other West African countries and trade in these food products occurs across borders of countries in the region. Cassava, when efficiently managed therefore, would not only improve the food security of Ghana but also that of the entire West African region.

Nweke et al. (1999) reported in their working paper on the Collaborative Study of Cassava in Africa (COSCA) that cassava is the most important of all the arable crops cultivated in Ghana. The COSCA report further revealed that villages in Ghana that did not experience famine in 1983 were those which cultivated cassava as the dominant staple crop. In contrast, areas where other major staples such as plantain, maize, millet and sorghum were considered to be more important experienced severe famine.

E. Moses (✉)
CSIR-Crops Research Institute, P.O. Box 3785, Kumasi, Ghana
e-mail: e.moses@cropsresearch.org

R.N. Strange and M.L. Gullino (eds.), *The Role of Plant Pathology in Food Safety and Food Security,* Plant Pathology in the 21st Century 3,
DOI 10.1007/978-1-4020-8932-9_2, © Springer Science+Business Media B.V. 2010

Famine rarely occurs in areas where cassava is grown widely because of its relative reliability compared with other crops. This may be a consequence of its greater adaptability to a range of climatic and edaphic conditions, including tolerance to drought, some pests and diseases (Romanoff and Lynam 1992; Fresco 1993).

The mean yield in Ghana of cassava root on a fresh weight basis is 13.0 t/ha (Nweke et al. 1999) although yields as high as 30.0 t/ha or more can be achieved. Diseases are among the major constraints that prevent achievement of optimum yields. The incidence and severity of African Cassava Mosaic Disease (ACMD) and Cassava Anthracnose Disease (CAD) are high or very high in cassava growing regions of Ghana (Moses and Lamptey 2001) and most of the cassava varieties cultivated, including improved varieties that have been introduced, are susceptible to these two diseases. ACMD, characterized by mosaic leaf development and stunting (Fig. 2.1), alone can cause yield losses as high as 50% in susceptible cultivars (Fauquet and Fargette 1990). CAD is characterized by cankers or wounds on stems of infected plants (Fig. 2.2). The disease is capable of causing losses of 30% or more in susceptible cultivars and affects the availability of planting materials.

The incidence of Cassava Bacterial Blight (CBB) caused by *Xanthomonas campestris* pv. manihotis is increasing in Ghana, probably due to expansion in cassava cultivation to feed emerging starch producing industries. CBB causes characteristic leaf blight and dieback symptoms (Fig. 2.2) that may result in total crop failure and famine such as occurred in Nigeria and Zaire (Democratic Republic of the Congo) in the early 1970s if conditions are favourable for the disease (Williams et al. 1973; Lozano 1986). Moreover, "candlestick plants" that results from severe CBB attacks may create shortages in the supply of planting material.

In recent years, a root rot caused by the basidiomycete, *Polyporus sulphureus*, has become the most threatening disease of cassava in Ghana. This disease was not

Fig. 2.1 Cassava showing symptoms of African Cassava Mosaic Disease (*left*) and a healthy plant (*right*)

Fig. 2.2 Cankers on the stems of a susceptible cultivar of cassava caused by cassava anthracnose disease (CAD, *left*) and symptoms of cassava bacterial blight (CBB, *right*)

Fig. 2.3 Rotten roots of cassava caused by *Polyporus sulphureus*. Note the large yellow fruiting body of the fungus (*left*). On the *right*, a cassava variety susceptible to bud necrosis disease showing dark necrotic areas symptomatic of the disease

mentioned in the COSCA studies conducted in the late 1980s indicating that farmers did not observe this fungus as a parasite of cassava before or during that period (Nweke et al. 1999). The pathogen causes rots on susceptible cultivars (Fig. 2.3) that may result in 100% loss of edible roots. In Sabado and Aveme farming communities of the Kpando District of the Volta region, *Polyporus* root rot is endemic with incidences as high as 80% or more recorded in surveys in 2000, forcing some farmers to stop cultivating the crop. Losses have also been recorded in the Gomoa and Swedru districts of the Central Region and the fungus also has a narrow distribution in the Ejura Sekyere Dumase district of the Ashanti region.

Despite the conspicuous presence of diseases and their effects on production, the majority of farmers in Ghana and most parts of Africa do very little or nothing at all to control diseases owing to lack of knowledge. African cassava mosaic disease for example is so prevalent in some countries that the leaf symptoms are often regarded as normal features of the plant (Thresh et al., 1994a, 1997). Awareness of the importance of cassava diseases in farming communities in most parts of Africa is low and it is common to find farmers in high disease pressure areas adopting and growing varieties that give just satisfactory yields when infection levels are clearly very high (Thresh et al., 1994b).

There is an increase in demand in Ghana for cassava as a food for human consumption because of the increasing human population and as an industrial raw material, particularly for starch production. Ghana has reached a point where uncontrolled losses in root and leaf yields due to diseases is unacceptable if sustainable food security for its people is to be achieved. This statement may apply to several sub-Saharan African countries that depend on cassava as a food security crop. The recent big increases in prices of food and shortages in the supply of some major staples in some parts of the world dictate that a greater effort is required, particularly by third world countries, to improve their food production.

This paper reports on strategic activities developed and conducted under the ISPP Congress Challenge Award Project of 2003 to control the major diseases of cassava in Ghana, including *Polyporus* root rot in the endemic areas of the Kpando District.

Varieties were tested for resistance, particularly to ACMD and *Polyporus* root rot, because of the threat the two diseases pose to cassava production. Also the knowledge and skills of farmers and agriculture extension agents in disease identification and control, which are important strategies required for any efficient disease control programme, were enhanced by field days and workshops.

2.2 Materials and Methods

2.2.1 Surveys to Establish the Incidence of Diseases

Surveys were conducted in 2000, 2004 and 2006 in the Sabadu and Aveme cassava growing areas of the Kpando District (the study communities) to establish the incidence and severity of major cassava diseases, including *Polyporus* root rot. The survey in 2000 was conducted primarily to identify *Polyporus* root rot hot spots. In each of the years twenty farms were randomly screened and on each farm, thirty plants were randomly examined. Disease severity was scored on a 1–5 scale (1 = no symptoms of disease observed; 5 = symptoms showing irreparable damage to organs). Plants with fruiting bodies of *P. sulphureus* attached were uprooted in order to measure the level of root rot compared to roots of healthy plants. Farmer practices that promoted the spread of diseases were documented.

2.2.2 Testing Cassava Genotypes for Resistance to Diseases

Nine elite genotypes of cassava from CSIR-Crops Research Institute's Cassava Breeding Programme, which have exhibited high tolerance to major cassava diseases such as ACMD, CAD and CBB in multi-locational trials in agro-ecologies similar to that of the two study communities, Sabadu and Aveme, were evaluated in the disease hotspots identified in the surveys. The nine genotypes were 96/0160, 96/0603, 96/1642, 96/1565, 96/1569, 97/4962, 97/3982, 97/4414 and 'Afisiafi'. The popular farmers' local cultivar was included as a control. Stem cuttings (30 cm in length) were planted 1 m apart. Weeds were controlled manually using hoes and cutlasses. Trials were established in August 2004 and diseases were assessed monthly beginning 3 months after planting on a 1–5 scale as above. Thirty plants of each genotype were assessed. In addition, the reactions of the cassava genotypes to major insect pests such as cassava green mites (CGM) which were prevalent in the area were recorded. Plants were harvested 15 months after planting.

2.2.3 Farmer Field Schools (FFS)

Two farmer field schools (FFSs), with 30 student farmers each, were established and operated in the two study communities, the disease resistance trials (see Section 2.2) being used as training fields. Timing of school sessions was based on the life cycle of the cassava plant to ensure that special problems associated with the various stages of growth were covered. They consisted of site selection and land preparation, selection of healthy planting material and improvement of cultivation practices, training in disease identification and control, harvesting and post-harvest disease management. At each session, farmers' experiences were first shared before resource persons introduced improved methods. Classroom discussions were followed by field work. Field days, when farmers demonstrated the knowledge and skills gained, were part of FFS activities.

2.2.4 Awareness and Disease Control Workshops

Disease awareness and control workshops were organized in strategic districts which cultivate cassava on a large scale and have reported cases of high disease incidence.

Workshops were conducted in four major cassava producing districts which had reported cases of *Polyporus* root rot or were threatened with the disease from neighbouring farms and had high incidence of ACMD.

Farmers who participated in the workshops actively cultivated cassava and had high incidences of *Polyporus* root rot or were threatened by occurrence of the disease in neighbouring districts or farms. Agriculture extension agents who participated in the workshops were drawn from major cassava producing areas where high disease pressures occurred. Workshop participants were trained in disease identification and control practices.

2.2.5 Introduction of Sweetpotato Varieties into Areas Where Cassava Diseases Are Endemic

To reduce overdependence on cassava in the study communities where cassava disease pressure was high, five high yielding varieties of sweetpotato, 'Faara', 'CRI-Otoo', 'CRI Apomoden', 'CRI-Ogyefo' and 'Sauti', which were pest and disease resistant, were introduced to farmers through the field schools. They were planted in demonstration plots and farmers were encouraged to select vines of varieties of their choice at harvest for planting on their own farms.

2.2.6 Multiplication Fields for Healthy Planting

Multiplication fields for healthy planting material for FFS members were established at Aveme using the three high yielding pest and disease resistant cassava varieties: 'CRI-Doku Duade', 'CRI-Esam Bankye' and 'CRI-Bankye Hemaa' from CSIR-Crops Research Institute's Cassava Breeding Programme. These varieties are known to have resistance to ACMD, CAD and did not show symptoms of attack by *P. sulphureus* during two seasons, when planted in hot spot areas for the fungus.

2.2.7 Development of Disease Identification and Control Extension Materials

Field research was conducted in hot spot areas for cassava diseases in order to provide material for the writing of an extension guide and the production of a Digital Video Disc (DVD) on disease identification and control. The book and the DVD, aimed at extension agents, farmers and students, have been produced and distribution of these materials to target groups is on-going.

2.3 Results

2.3.1 Surveys to Establish Disease Incidence

First contact with the two communities of Sabadu and Aveme was made in 2000. Farmers were immediately made aware of cassava diseases and were introduced to improved cultural practices that reduce disease persistence and spread. Incidence of *Polyporus* root rot in the study area in 2000 was 85% whereas incidences in 2004 and 2006 were 44% and 38%, respectively, indicating a decline in the number of farms having the disease.

Other major diseases of cassava identified in the survey were ACMD, CAD and Bud Necrosis Disease (BND) (Figs. 2.1–2.3). Incidences of ACMD, CAD and BND in 2004 were 100%, 60% and 30%, respectively. A disease severity of 4.2 on a scale of 1–5 was recorded for ACMD in the study community in 2004. Severity of CAD and BND were 3.3 and 3.2 respectively on a 1–5 scale. Farmers in the two communities cultivated their local variety.

2.3.2 Testing Cassava Genotypes for Resistance to Diseases

Reactions of the varieties to various diseases of cassava in the disease resistance trials are shown in Table 2.1. The farmers' popular local variety was severely affected by ACMD, scoring 4.2 on a 1–5 scale. The introduced varieties did not show any symptoms of ACMD 15 months after planting. None of the varieties was affected by *Polyporus* root rot in the 15 month period, although the farmers' local variety is known to be susceptible. The fruiting bodies of the pathogen were, however, observed in the area of the trial. Two of the introduced test varieties, 96/1569 and 97/4962 were affected by BND at a disease severity of 4.2 and 3.8 respectively. The two varieties, 'Afisiafi' and 96/1569 were affected by CAD at severity levels of 2.3 and 3.2 respectively. Four of the introduced varieties, 96/0160, 96/0603, 97/3982 and Afisiafi were susceptible to cassava green mites (CGM) at severity levels of 3.1, 2.8, 2.3 and 2.2 respectively. The variety 97/4962 was susceptible to brown leaf spot disease (BLS) with a score of 2.3.

With the exception of the two varieties 96/1565 and 96/1569, the other introduced varieties gave root yields higher than the farmers' local cultivar of 11.0 t/ha. Three of the introduced varieties, 'Afisiafi', 96/1642 and 96/0160 gave root yields of 24.5, 23.0 and 22.0 t/ha, respectively, double that of the farmers' local variety (Fig. 2.4). The quality of the roots was good even at harvest 15 months after planting (Fig. 2.5).

Table 2.1 Reaction of cassava genotypes to diseases, cassava green mite and root yields in a trial

Variety	*Polyporus* root rot	ACMD	CBB	CAD	BLS	Bud necrosis	CGM	Yield (t/ha)
96/0160	1.0	1.0	1.0	1.0	1.0	1.0	3.1	22.0
97/4962	1.0	1.0	1.0	1.0	2.3	3.8	1.0	19.3
96/1642	1.0	1.0	1.0	1.0	1.0	1.0	1.0	23.0
96/0603	1.0	1.0	1.0	1.0	1.0	1.0	2.8	20.0
Local	1.0	4.2	1.0	1.0	1.0	1.0	1.0	11.0
97/3982	1.0	1.0	1.0	1.0	1.0	1.0	2.3	20.5
Afisiafi	1.0	1.0	1.0	2.3	1.0	1.0	2.2	24.5
97/4414	1.0	1.0	1.0	1.0	1.0	1.0	1.0	16.2
96/1565	1.0	1.0	1.0	1.0	1.0	1.0	1.0	6.5
96/1569	1.0	1.0	1.0	3.2	1.0	4.2	1.0	7.8

ACMD, African cassava mosaic disease; CBB, cassava bacterial blight; CAD, cassava anthracnose disease; BLS, brown leaf spot; CGM, cassava green mite. Plants were scored on a 1–5 scale where 1 represents no disease and 5 represents irreparable damage to organs.

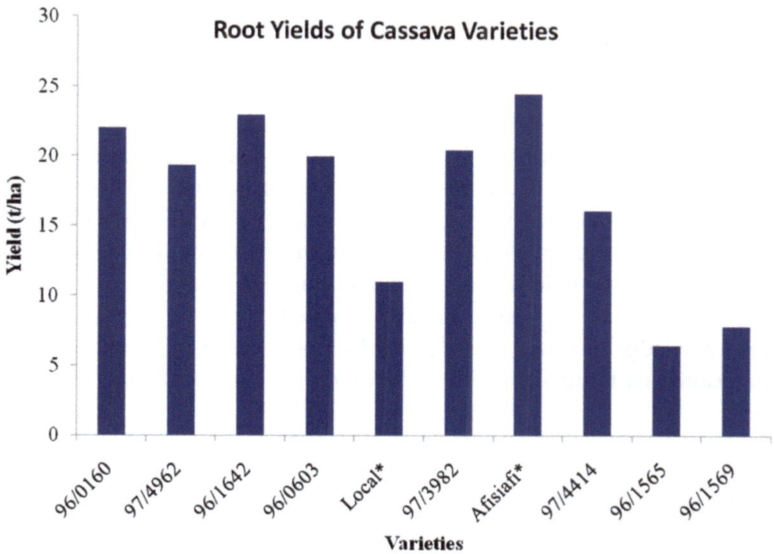

Fig. 2.4 Yield of cassava varieties screened for disease resistance

Fig. 2.5 Farmers showing good quality edible roots of variety 96/1642 at harvest 15 months after planting (*left*) and a disease control workshop at Kpando (*right*)

2.3.3 Farmer Field Schools (FFS)

Sixty farmers registered and participated in the farmer field schools (Fig. 2.5) where, during the 3 year period of the project they became skilled in site selection, land preparation, pre-season management of diseases on-farm, selection of healthy planting materials and good planting methods for cassava and sweet potato. They developed knowledge and skills in the identification and control of all major diseases of cassava prevalent in the study area and were introduced to improved cultural methods and helped to put them into practice on their own. Removal of young fruiting bodies of *P. suphureus* immediately they appeared and destroying them by burning was emphasized as an improved cultural practice that must be extended to neighbours who did not attend the schools. Regular field walks for early detection of diseases, roguing and maintenance of good farm sanitation were other areas that were emphasized at the field schools. The importance of rotating cereals and grains with cassava after every 3 years was also stressed. Farmers were encouraged to share the knowledge obtained from the field schools with neighbouring farmers in order to ensure that farms surrounding their own were not sources of inoculum.

2.3.4 Workshops for Disease Awareness and Control

Two hundred farmers and agricultural extension agents in four major cassava growing districts in southern Ghana: Ejura Sekyere Dumase, Gomoa, Kpando and Hohoe have benefited from the workshops designed to increase disease awareness and disease control. Experiences of diseases affecting cassava production on their farms and in their operational areas were discussed and solutions to disease problems proposed.

The Workshops focused on disease identification, based on easily recognized symptoms, integrated disease management practices, that is the use of disease resistant varieties, healthy planting materials and improved cultural practices, including skills in sanitation practices that reduce disease levels on farms. The importance of local quarantine measures to reduce the spread of *Polyporus* root rot into new localities was emphasized.

English and local dialects specific to the districts concerned were used to facilitate understanding at the workshops. Collaboration with the Ministry of Food and Agriculture Directorates of the four districts contributed to the success of the workshops.

2.3.5 Introduction of Sweetpotato Varieties to Disease Endemic Areas

A good harvest of the five sweetpotato varieties, 'Faara', 'CRI-Otoo', 'CRI Apomoden', 'CRI-Ogyefo' and 'Sauti' was obtained 4.5 months after planting the demonstration plots. Yield characteristics of the five varieties from evaluation studies in agro-ecologies similar to that of Sabadu ranged from 13 to 35 t/ha of tubers (Table 2.2) and tuber characteristics of two of the introduced sweet potato varieties are shown in Fig. 2.6. Farmers selected vines of their desired varieties for planting on their own farms.

Table 2.2 Yields of introduced sweet potato varieties

Variety	Yield range (t/ha)
CRI-Otoo	15–23
CRI-Apomoden	20–35
'Sauti'	13–20
'Faara'	15–20
CRI-Ogyefo	15–20

Fig. 2.6 Section through the tuber of sweetpotato variety 'CRI Apomoden' (*left*) and section through the tuber of sweetpotato variety Faara (*right*)

2.3.6 Development of Disease Identification and Control Extension Materials

A 32 page booklet entitled *Guide to Identification and Control of Cassava Diseases* was written and published as a major output of the ISPP 2003 Congress Challenge Project. Several copies of the book have been distributed to farmers and extension agents of the Ministry of Food and Agriculture. The book has been distributed to libraries of second cycle and tertiary institutions including Agricultural Faculties of Universities in Ghana. Symptoms of all the major diseases, such as African Cassava Mosaic Disease (ACMD) and Cassava Bacterial Blight (CBB) and the major root rot diseases in Ghana including rots caused by *P. sulphureus* are well described. Control measures specific for each disease are documented and a full copy of the booklet has been posted at the Website of the ISPP.

In addition, a DVD on cassava diseases was developed and produced as extension material for farmers, extension agents and students of agriculture. Symptoms of all the major diseases are illustrated with full colour pictures and described, making identification an easy exercise. Control measures that can be applied for specific diseases are also described. A summary of facts on disease control strategies in a major Ghanaian language, 'Twi' spoken by 70% of the country's population is included on the DVD to help illiterate farmers benefit from this material.

2.4 Discussion

The threat of diseases of cassava to sustainable food security in Africa is real as demonstrated by the recent pandemic of ACMD in East Africa and the resultant food shortages in some parts of that sub-region (Otim-Nape et al. 1997). As the majority of farmers in Ghana do very little or nothing to control diseases on their farms, losses caused by plant pathogens are very common in all crop production systems. Close observation of farmers and extension agents in participatory research activities revealed that lack of knowledge or poor understanding of diseases may be responsible for the inaction of the two groups. Several farmers in cassava growing regions of Africa regard symptoms of cassava mosaic disease, for example, as a normal feature of the plant (Thresh et al. 1994a, 1997). This is unacceptable on a continent where cassava is one of the major food security crops. Increased awareness by farmers, extension officers and other stakeholders of diseases and their importance is necessary if any meaningful improvement in yields is to be achieved through disease control.

The activities of the 200 farmers and extension agents trained in disease identification and control may be contributing to the reduction of incidence and severity of diseases of cassava in the four districts described in this report. In particular, the activities of the 60 farmers of the Farmer Field Schools, operated in areas where *Polyporus* root rot is endemic, may be contributing towards the reduction in the incidence of this disease. Ghana has more than one hundred districts that cultivate cassava actively under high disease pressure. It is therefore, necessary for the training of farmers and extension agents reported in this paper to be repeated in other parts of the country. A nationwide campaign to improve disease control is now more important than ever as there is a determination in Ghana to make cassava an important industrial raw material. More cassava will have to be produced in order to reduce competition between human and industrial demands and farmers and extension agents will have to be better trained in order to combat the greater disease problems that are likely to arise from more intensive cultivation of the crop.

Development and deployment of disease resistant varieties for the control of cassava diseases will continue to be the most effective methods of reducing losses in yields (Otim-Nape et al. 1997). Disease resistance when identified is often long lasting and friendly to the environment. It is encouraging therefore, that three of the varieties 96/0160, 96/1642 and 'Afisiafi' each produced root yields higher than 22.0 t/ha compared with a yield of 11.0 t/ha from the farmers' popular local variety. These three varieties also exhibited high resistance to ACMD compared to the farmers' variety. Farmers in the two communities now have increased access to high yielding pest and disease resistant varieties. Effective management of these introduced varieties could double production in areas of high disease pressure, such as the Kpando district, improving both food security and incomes.

The introduction of five high yielding pest and disease resistant sweet potato varieties was welcomed by farmers who eagerly took vines from the demonstration fields for planting on their own farms. Adoption of sweet potato in the Sabadu and Aveme communities will reduce their overdependence on cassava. Moreover, four of the varieties have high market demand in urban communities in the Volta region and therefore a good capacity for income generation. They can be harvested 4.5 months after planting, giving minimum yields of about 13 t/ha. This can be doubled if two crops are produced annually and would consequently compensate for not cultivating cassava. Additionally, the value of sweet potatoes weight for weight is greater than that of cassava in all parts of Ghana. Farmers who grow sweet potato in addition to cassava will surely improve their annual incomes and food security compared with those who grow cassava alone.

The sweet potato variety CRI-Apomoden is orange fleshed with high beta carotene levels (Fig. 2.6). Annually in Ghana the Ministry of Health administers vitamin A to infants to help prevent vitamin A deficiency among children. The introduction of orange fleshed sweetpotato into the community is aimed at improving sources of vitamin A through nutrition.

In Ghana and most parts of sub-Saharan Africa, prices of food continue to rise alongside reports of food shortages. Field and post-harvest losses in yields of major staples, such as cassava, to uncontrolled diseases are unacceptably high. Food insecurity will persist in most parts of the continent if such diseases are not controlled. However, although disease control could help to increase crop yields substantially, they are insufficient on their own to ensure a good level of sustainable production. Improvements in agronomic practices such as promotion of soil fertility, water collection and its storage for irrigation and post-harvest management of farm produce will also be required. In some parts of Africa, the collection of rainfall and its careful use in irrigation is particularly vital as, without it, little of the crop would survive, making other good agronomic practices redundant. It is important that policy makers who determine the support given to Agriculture in different parts of Africa consider food security as a function of the implementation of integrated agronomic systems which include those described in this paper.

References

Fauquet C, Fargette D (1990) African cassava mosaic virus: etiology, epidemiology and control. Plant Dis 74:404–411

Fresco LO (1993) The dynamics of cassava in Africa, an outline of research issues. COSCA Working Paper No. 9. Collaborative Study of Cassava in Africa. IITA, Ibadan, Nigeria

Lozano JC (1986) Cassava bacterial blight: a manageable disease. Plant Dis 70:1089–1093

Moses E, Lamptey JNL (2001) Cassava diseases in Ghana. Annual Report of CSIR Crops Research Institute, Ghana

Nweke FI, Haleegoah J, Dixon AGO, Ajobo O, Ugwu BO, Al-Hassan R (1999) Cassava production in Ghana. A function of market demand and access to improved production and processing technologies. Collaborative Study of Cassava in Africa. COSCA Working Paper No. 21. IITA, Ibadan, Nigeria

Otim-Nape GW, Thresh JM, Bua A, Baguma S, Ogwal S, Semakula GN, Acola G, Byabakama B, Martin A (1997) Cassava mosaic virus disease in Uganda. Hobbs Printers, Hampshire, UK, pp 3–5; 57–60

Romanoff S, Lynam J (1992) Commentary. Cassava and African food security: some ethnographic examples. Ecol Food Nutr 27:29–41

Thresh JM, Fishpool LDC, Otim-Nape GW, Fargette DA (1994a) African cassava mosaic virus disease – an underestimated and unresolved problem. Trop Sci 34:3–14

Thresh JM, Otim-Nape GW, Jennings DL (1994a) Exploiting resistance to African cassava mosaic virus. Aspects Appl Biol 39:5–60

Thresh JM, Otim-Nape GW, Legg JP, Fargette D (1997) African cassava mosaic virus disease: the magnitude of the problem. Afr J Root Tuber Crops 2:13–19

Williams RJ, Agboola SD, Schneider RW (1973) Bacterial wilt of cassava in Nigeria. Plant Dis Rep 57:824–827

Chapter 3
Biosecurity in the Movement of Commodities as a Component of Global Food Security

N.A. van der Graaff and W. Khoury

3.1 Introduction

The introduction of agents of plant diseases, other plant pests[1] and emerging plant pests are part of a larger group of similar subjects that are covered within the term biosecurity. Biosecurity is used in this paper in the FAO sense (FAO 2007a): Biosecurity is a strategic and integrated approach that encompasses the policy and regulatory frameworks (including instruments and activities) for analyzing and managing relevant risks to human, animal and plant life and health, and associated risks to the environment. Biosecurity covers food safety, zoonoses, the introduction of animal and plant diseases and pests, the introduction and release of living modified organisms (LMOs) and their products (e.g. genetically modified organisms or GMOs), and the introduction and management of invasive alien species.

The interest in biosecurity has risen over the last decade in parallel with increasing trade in food, plant and animal products, growing international travel, heightening awareness of biological diversity and the impact of agriculture on environmental sustainability.

Biosecurity in relation to plant disease covers the prevention of the introduction of plant diseases and aspects negatively affecting food safety, that is pesticide residues in food and mycotoxins in food. This paper primarily covers the prevention of the introduction of new plant diseases, its relation to food security and the national and international technical and policy framework to address plant biosecurity.

N.A. van der Graaff (✉)
Via G. Viotti 31, 00124, Rome, Italy
e-mail: vdgraaff@alice.it

W. Khoury
Food and Agriculture Organization of the United Nations, Via Delle Terme di Caracalla, 0100, Rome, Italy

[1]Pest: "Pest" – any species, strain or biotype of plant, animal or pathogenic agent injurious to plants or plant products (IPPC 2007). This definition includes weeds.

R.N. Strange and M.L. Gullino (eds.), *The Role of Plant Pathology in Food Safety and Food Security,* Plant Pathology in the 21st Century 3,
DOI 10.1007/978-1-4020-8932-9_3, © Springer Science+Business Media B.V. 2010

3.2 The Introduction of Plant Pests

Many plant pests do not occupy the whole ecological region suitable to them due to physical obstacles (oceans, seas, mountain ranges, deserts) to their spread. Thus, other geographical regions may hold plant species that are susceptible to a pest but are not exposed to it. Three situations exist:

1. A particular host/pest association may have evolved in one geographical region, but other, often related, plant species evolved in other geographic regions may show susceptibility if they encounter the pest. When a species is introduced into a new area, it may acquire pests not known before on that species; similarly, pests originating in one region, may find new suitable host species after introduction into other regions.

   For example, *Coffea arabica* probably encountered an unimportant pathogen of a related species (*Coffea eugenioides*?) when *C. arabica* cultivation was extended to Western Kenya in the 1920s (Robinson 1976). The pathogen, *Colletotrichum kahawae*, the cause of Coffee Berry Disease, subsequently spread over all *C. arabica* growing regions of Africa, causing important economic damage. Similarly, although six species of *Sclerospora* and two species of *Sclerophthora* cause downy mildew of maize, none of these originate from the Americas, the centre of origin of maize (Frederiksen and Renfro 1977); and Cassava Mosaic Virus, a serious disease in Africa, does not occur in South America the centre of origin of cassava.
2. Man has introduced many crop plants into areas outside their centres of origin and in this process many pests were left behind. For example, all crop plants were introduced into Australia; and many of their pests do not occur in that country. Rubber was introduced into Asia without South American leaf blight, *Microcyclus ulei*, which made large scale cultivation of rubber in South East Asia possible. Potatoes were taken outside South America without *Phytophtora infestans*, which, after its introduction, caused the great Irish potato famine.
3. When cropping systems and crops change, new, more aggressive, genotypes of pathogens may emerge. For example, Enset wilt probably evolved to a major wilt disease of banana (*Xanthomonas campestris pv. musacearum*), which is now widely spread over East and Central Africa (Tushemereirwe et al. 2004). Similarly, the East African Cassava Mosaic Virus (EACMV), a new strain of Cassava Mosaic Virus, evolved probably in Uganda and subsequently spread widely through East Africa (Legg and Tresh 2000).

Many introductions of plant pests have caused substantial losses. A well known example is the introduction of the insect pest *Phylloxera (Daktulosphaira vitifoliae)* on grapes in Europe in the 19th century, which led to (1) the replacement of all grapes in Europe with planting material grafted on resistant rootstocks and (2) the first international plant health treaty, the International Phylloxera Convention, which was signed in Bern in 1881. The use of American rootstocks resistant to *Phylloxera* resulted in the additional introduction of grape downy mildew in Europe and the consequent discovery of Bordeaux mixture.

A few examples of plant diseases spreading over the latest 30 years include: coffee leaf rust, a disease which spread in some 100 years from Africa over practically the whole world; the debilitating citrus greening disease (Asian form), which gradually infested all Asian countries and has now reached Brazil and Florida; the brown citrus aphid, the efficient vector of citrus tristeza virus, spread from South America to Central America and the Caribbean causing massive outbreaks of citrus tristeza requiring a change of rootstock and, thus, the replacement of practically all citrus in those countries[2]; black Sigatoka of banana; bayoud disease of the date palm; and lethal yellowing of coconut palms.

Recent examples of major introductions with substantial financial consequences are soybean rust into the Americas, while the emerging virulent races of yellow and black rusts of wheat (YR9 and Ug99) raise the possibility of a global pandemic. Introductions are not limited to agricultural crops and ornamentals only; the introduction of pests, including diseases, of forests and the natural flora as a whole are common and difficult to control.

Global trade and distribution systems limit the chance that new introductions and newly emerging plant diseases will result in large scale hunger. However, many introductions continue to occur and cause serious damage. The enormous increases in trade and international travel substantially heighten the risk of new introductions. On the other hand, phytosanitary measures, taken by Governments to protect their crops and native flora from pest introductions are considered to be major trade barriers. With the decline of other trade barriers over the last 18 years, the interest in, so called, 'Sanitary and Phytosanitary' (SPS) measures[3] has substantially increased, and concern about their misuse as trade barriers resulted in the conclusion of the Sanitary and Phytosanitary agreement as part of the Uruguay Round of trade negotiations (WTO 2007a).

---

[2]Tristeza can be contained in the absence of the efficient vector. The Meditteranean is one of the few areas remaining substantially free of Tristeza, and where practically all citrus is planted on susceptible rootstock. Unfortunately, the efficient vector of Tristeza was found in Portugal in 2003.

[3]*Measures*: In the literature there is often confusion of the meaning of Measures, Regulations and Standards. In Sanitary and Phytosanitary word use, Measures and Regulations are often used interchangeably. They are issued by individual Governments and are mandatory in the national territory. Sanitary and Phytosanitary (SPS) Measures are potential technical barriers to trade, and the WTO SPS agreement stipulates therefore that measures be scientifically justified, and have the least effect on trade.

International and regional standards: Are developed by the members of intergovernmental regional and international organizations as guidance for the SPS measures of governments and are voluntary. (If they were to be binding, they would have the nature of an international treaty or the extension of an International Treaty).

Private standards: are standards set by groups of producers, traders, retailers or others and detail the conditions their produce or merchandise should meet. Retail through supermarkets has increased tremendously and their procurement systems are changing traditional trade. These changes have great effects on international trade. Modern purchase systems often include private standards that address the supply chain from farm to fork. Private standards address food safety, but also other issues like labour, health, safety and wellbeing, environmental concerns, and sustainability; phytosanitary issues are, in general, not part of these standards. Private standards are voluntary, but their conditionality for market access enforces their use.

3.3 Food Security and Plant Diseases

The world population continues to increase albeit at a gradually declining rate. It is now expected that the world population may stabilize around 2050. According to the UN world population prospects (UN 2006) the world population will likely increase from the current 6.7 billion to 9.2 billion in 2050.

It is expected that demand for food and other agricultural produce will continue to increase over this period but at a much slower rate than in the past century. Demand for other uses of agricultural produce, such as biofuel and animal feed, will contribute to this rise in demand (Bruinsma 2003). Food production will also be influenced by climate change, in particular in irrigated areas, where production is expected to drop. The use of corn and oil crops for biofuel has contributed to the increases in food prices in 2007/2008. The estimates of the increases in food price due to the use of crops for biofuel production vary widely.

Food security was defined by the World Food Summit in 1996 as 'When all people, at all times, have physical, social and economic access to sufficient, safe and nutritious food which meets their dietary needs and food preferences for an active and healthy life'. Components of food security are food availability, food access, food utilization, (and food stability).

Availability of Food Plant diseases are part of the spectrum of plant pests that reduce food production. Indigenous plant pests, the entry, establishment and spread, and the emergence of plant diseases result in yield reductions of food crops. Overall quantifications of losses and potential losses due to plant pests are available but notoriously unreliable.

Access to Food Most people in the world do not produce food themselves and, thus, other activities have to provide them with income to procure food. National and international agricultural trade is of great importance for ensuring access to food and international trade has increased very substantially over the latest decades. 'Few countries could survive the elimination of agricultural trade without a considerable drop in national income, and none could do so without considerable reduction in consumer choice and well-being' (Bruinsma 2003). Plant pests directly reduce access to food through reduction of income from yields of food and cash crops, reduction in forest productivity, as well as increased costs of control. Indirect effects are reduced access to international markets due to the inability of meeting phytosanitary regulations of importing countries or due to the cost of complying with such regulations.

Utilization of Food Safe and nutritious food is an important factor of food security. To protect the health of the population, governments set maximum residue levels of pesticides and mycotoxins in food; the implementation of these measures has direct effects on national and international trade in food.

3.4 World Trade, International Travel and Their and Effects on Biosecurity

World trade in agricultural products has changed enormously since 1960. The possibilities for exports and imports of fresh and processed products have greatly increased with improved transport possibilities, in particular air transport. Since the early 1960s, the nominal value of agricultural exports has increased tenfold. In the early 1960s, developing countries had an overall agricultural trade surplus of almost US$7 billion per year. By the end of the1980s, however, this surplus had disappeared. During most of the 1990s and early 2000s, developing countries were net importers of agricultural products, and the least developing countries shifted from net exporters to net importers (FAO 2007b). Both developing and developed countries are importing more high value and processed foods, among which are fruits and vegetables. Driving factors for the increase in this demand include rising incomes, urbanization, trade liberalization, investment and advancing technology (Worldbank 2007). Improving market infrastructure will continue to enhance opportunities to the rural poor for increased trade.

International travel has tremendously increased over the last decades, with air transport taking over from other means of transport. International business travel and, in particular, private travel and tourism provide new pathways to introduction of hazardous pests into new areas. A number of countries take measures to contain risks of introduction of plant pests through international travel (passenger inspection, aircraft treatment); others do not seem to consider international travel as a major pathway for introductions and therefore take a less strict attitude. A quantification of the effects of these different approaches is not available.

3.5 The National and International Plant Health Biosecurity Framework

Plant health biosecurity has national, regional and international dimensions. The primary responsibility of plant health biosecurity is national, but pests of quarantine concern are transboundary and therefore need also regional and international action. Only a solid national infrastructure makes it possible for countries to cooperate on regional and global levels.

3.5.1 *National Plant Health Systems*

3.5.1.1 National Phytosanitary Measures

Biosecurity concerns primarily national regulatory systems. Biosecurity measures relevant to plants and plant products are (1) phytosanitary measures[4] and (2) food safety measures as far as they relate to pesticides and mycotoxins in food.[5] To protect their agriculture and national resources (i.e. biodiversity and the natural flora), governments take phytosanitary measures to prevent the entry and establishment of plant pests, including plant disease causing organisms. The objective of those measures is (1) to reduce the risk of entry, establishment or spread of quarantine pests[6] whose introduction and/or spread will result in economic losses and could require expensive eradication or control operations and (2) to reduce the impact of regulated non-quarantine pest[7] (for example, if national nurseries are only allowed to produce virus free plants, imported plants of the same species must meet the same criterion).

In this objective, environmental damage (i.e. losses to plant biodiversity and the natural flora) would be given an economic weight. Invasive Alien Species are plant pests if they threaten plant biological diversity; Living (Genetically) Modified Organisms (LMOs) are equally plant pests if they threaten crops or the native flora (relevant LMOs will mostly be modified plants and will be plant pests if they express weedy characteristics).

While Phytosanitary Measures should be designed to meet the above objective, but their effect on the movement of goods and people should be limited to the necessary minimum.

3.5.1.2 The Organization of National Plant Protection

Governments establish, to the best of their ability, national plant protection services which are responsible for (FAO 1997):

---

[4]Any legislation, regulation or official procedure having the purpose to prevent the introduction and/or spread of quarantine pests, or to limit the economic impact of regulated non-quarantine pests (FAO 1997).

[5]Food safety measures: Any measure to protect human or animal life or health from risks arising from additives, contaminants, toxins or disease-causing organisms in foods, beverages or feedstuffs and to protect human life or health from risks arising from diseases carried by plants or products thereof (WTO 2007, SPS agreement adapted).

[6]A pest of potential economic importance to the area endangered thereby and not yet present there, or present but not widely distributed and being officially controlled (IPPC 2007).

[7]A pest whose presence in plants for planting affects the intended use of those plants with an economically unacceptable impact and which is therefore regulated within the territory of the importing contracting party (FAO 1997).

- The issuance of phytosanitary certificates
- The surveillance of growing plants, including both areas under cultivation and wild flora, and of plants and plant products in storage or in transportation
- The inspection of consignments of plants and plant products moving in international traffic
- The disinfestation or disinfection of consignments of plants, plant products and other regulated articles moving in international traffic
- The protection of endangered areas and the designation, maintenance and surveillance of pest free areas and areas of low pest prevalence
- The conduct of Pest Risk Analyses

Governments are also responsible for:

- The distribution of information within their territory regarding regulated pests and the means of their prevention and control
- Research and investigation in the field of plant protection
- The issue of phytosanitary regulations

These functions comprehensively cover the national requirements for the protection of plants and the requirements for import, export and international travel. Nevertheless, many developing countries and countries with small economies do not have the ability to provide comprehensively all these services.

Invasive Alien Species and Living Modified Organisms are often dealt with in other parts of the Government than plant health. Food safety and animal health are traditionally also addressed in other agencies. Additionally, Governments may have made provisions to deal with terrorist threats, either using these agencies or establishing new infrastructures. As a consequence, the national infrastructure dealing with biosecurity is often very fragmented. It should be noted, however, that a number of countries have merged agencies to make best use of synergies and resources among these subjects.

3.5.2 The International Regulatory Framework

Until the late 1980s, the *International Plant Protection Convention* (IPPC), an International Treaty adopted by the FAO Conference in 1951, was the only international agreement that dealt with phytosanitary issues in a global and comprehensive manner. Its main purpose is the securing of common and effective action to prevent the introduction and spread of pests of plants and plant products and to promote measures for their control. It covers both agricultural and non-agricultural plants (forests, natural flora). The Convention establishes joint responsibilities among parties to the Convention to take action at national, regional and global level, obliges parties to establish appropriate infrastructure for plant health, provides rights to parties to take phytosanitary measures, but limits the rights to those that can be economically and biologically justified, establishes a universal export certification procedure and provides for information exchange.

The *GATT agreement* of 1948 in its article 20 allowed for 'the adoption of enforcement of measures necessary to protect human, animal or plant life or health' provided that 'such measures are not applied in a manner which would constitute a means of arbitrary or unjustifiable discrimination between countries where the same conditions prevail, or a disguised restriction to international trade'.

The concern that sanitary and phytosanitary measures could be used as barriers to trade in an unjustified manner resulted in the negotiation and adoption in 1994 of the *WTO Agreement on the Application of Sanitary and Phytosanitary Measures* (SPS agreement) as part of the Uruguay round and as an implementation of GATT article 20. In the SPS agreement, it is recognized that countries can take SPS measures but that they should be scientifically based, and least interfering with trade. There are provisions for transparency, equivalence of measures, non-discrimination among national and international trade partners and the need for consistency of risk among measures (WTO 2007a).

Harmonizing SPS measures among countries reduces costs for the scientific justification of measures by importing countries and for meeting SPS measures of importing countries by their trade partners. Consequently the WTO SPS agreement recognizes standards, guidelines and recommendations developed by the *FAO/WHO Codex Alimentarius* for Food Safety, by the IPPC for Plant Health and by the *World Organization for Animal Health* (OIE) for Animal Health, and it strongly encourages the parties to base their SPS measures on these. Measures based on these standards, guidelines and recommendations are considered to meet the requirement of scientific justification. For ease of understanding, the Phytosanitary world has decided to use the word standard to cover all three terms used in the SPS agreement.

To make it possible to use the IPPC as the vehicle for harmonization, the FAO Conference in 1989, 1991, and 1993 took decisions to establish a standard setting mechanism and a secretariat. The standard setting programme started in 1992 and the first standard was adopted in 1995. The IPPC was amended by the FAO Conference in 1997 to include standard setting, to establish a Governing Body for the Convention (instead of the FAO Conference) and to formalize the Secretariat (for the history of IPPC till 1998, see [van der Graaff 1999]). The amendments came into force in 2005 for all parties.

The Parties to the Convention met annually from 1997 to 2004 as the Interim Commission on Phytosanitary Measures and since the amendments of 2005 came into force, they meet as the Commission on Phytosanitary Measures (CPM). Standards (International Standards for Phytosanitary Measures, ISPMs) are drafted by working groups, considered by a Standards Committee and adopted by the Commission. Most existing standards deal with concepts such as Pest Risk Analysis, pest free areas, pest free places of production etc. Recently, also commodity and pest specific standards are being developed, recognizing, however, that very few specific standards can have a world wide application. A specific ISPM with a world wide application is ISPM nr. 15: 'Guidelines for regulating wood packaging material in international trade'. The full text of ISPMs can be found on the IPPC website (https://www.ippc).

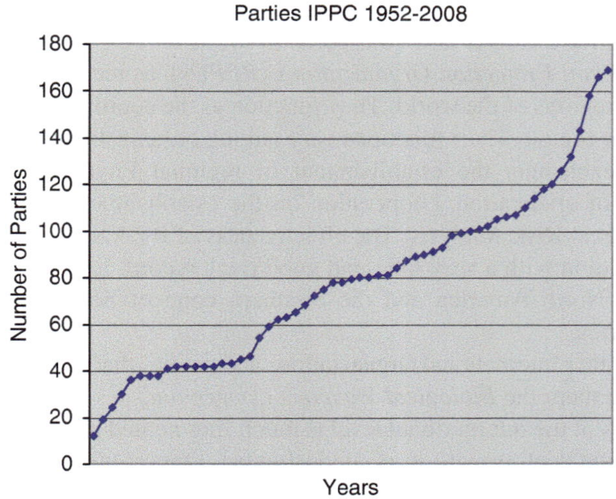

Fig. 3.1 Increase in the number of parties to the International Plant Protection Convention over the period 1951 to July 2008

The number of parties to the IPPC has increased from 93 in 1990 to 170 in August 2008, reflecting the increased importance countries give to the Treaty (Fig. 3.1). Wide participation in standard setting is achieved through expertise in working groups, regional representation in the Standards Committee, the opportunity for the provision of written comments and participation in the CPM. For example in 2008, 91 developing and 37 developed country parties participated in the third session of the CPM (FAO 2008a).

The *Convention on Biological Diversity (CBD)* was adopted in 1992 and came into force in 1993. Its objective is the conservation of biological diversity. It addresses in article 48 (h) the threat of Invasive Alien Species (IAS) to biological diversity (Secretariat to the Convention on Biological Diversity 2000a). The Conference of Parties adopted in 2002 the "Guiding principles for the prevention, introduction and mitigation of impacts of alien species that threaten ecosystems, habitats or species" (Convention on Biological Diversity 2002). As can easily be understood there is an overlap between the CBD and the IPPC, where it concerns Invasive Alien Species that are pests of plants, or biological control agents. Pest Risk Analysis for Invasive Alien Species that are plant pests is considered in the IPPC's ISPM nr. 11: "Pest Risk Analysis for quarantine pests including analysis of environmental risks and living modified organisms".

The *Cartagena Protocol on Biosafety to the Convention on Biological Diversity* was adopted in 2000 within the framework of the CBD and came into force in 2003. The Cartagena Protocol regulates international aspects of Living Modified Organisms (LMOs) (Secretariat to the Convention on Biological Diversity 2000b)

There is an overlap with the IPPC in cases where these LMOs are plant pests; Pest Risk Analysis for LMOs is also considered in ISPM nr. 11.

Regional Plant Protection Organizations (RPPOs), as recognized in the IPPC, exist in most regions of the world. They function as the coordinating phytosanitary bodies in their region. Their functions vary among organizations, but may include information exchange, the establishment of regional Phytosanitary standards, coordination of eradication, cooperation on the establishment of pest free areas and regional Pest Risk Analysis. The effectiveness of RPPOs varies. Europe has a large organization with a very long and good track record. Efficient organizations also exist in North America and the Southern cone of South America and in Central America.

There are other International Organizations and treaties that deal with Biosecurity issues, among them the *Biological Weapons Convention.*

Biosecurity at the international level is much fragmented and the Organizations and Secretariats dealing with it are underfunded. Opportunities for international action to support national systems cannot be followed up. The fragmentation has historical reasons and reflects in many ways the situation at the national level. Although a measure of consolidation at international level would in theory be preferable, in practice this is not easy as the existing treaties, organizations and programmes were concluded in different for after substantial negotiations followed by long adherence processes. Any change would need the political will of a large majority of countries. At present, cooperation and harmonization among the various entities appear to be the most promising course. In this respect, it should be noted that both for IAS and LMOs there is mutual recognition among the Governing Bodies of the IPPC, the CBD and the Cartagena Protocol and close cooperation among the Secretariats.

3.6 Trade Related Plant Biosecurity Issues Raised at the WTO SPS Committee

The WTO SPS Committee consists of WTO Members. It is responsible for managing SPS trade-related disputes and concerns among countries. WTO members are obliged to notify proposals for new SPS measures to the Secretariat of the WTO SPS Committee. The SPS Committee also receives reports from WTO members that have concerns about SPS measures taken by other countries (so called 'trade concerns'). Trade concerns raised by one or more countries might be supported by other Members that have the same concerns.

Solutions to trade concerns are sought but not always found, and issues may enter the dispute settlement procedure of the WTO. Between 1995 and 2008, a total of 71 trade concerns on plant health issues were raised in the SPS committee. Some concerned general issues and were not directed to a country or group of countries, eight led to some phase of dispute settlement and two went through a full dispute settlement procedure.

Overall data on trade concerns are regularly reported to the SPS Committee (WTO 2008) Tables 3.1–3.3 reflect those data. Phytosanitary measures against which trade concerns were expressed are provided in Table 3.1. Practically all of these phytosanitary measures were issued by developed countries and a limited number of mid-income developing countries. Phytosanitary concerns were mostly expressed by developed countries and by mid-income developing countries that are major exporters of agricultural produce. Some Least Developed Countries (LDCs) supported the trade concerns expressed by others (Tables 3.2 and 3.3). It may be

Table 3.1 WTO SPS Committee, 1995–2007 (WTO 2008): Phytosanitary measures against which trade concerns were expressed according to the region of the WTO member country that issued the measure

	Measures issued by WTO members in					
	Developed countries	Other member countries: Africa	Asia	Europe	Latin America	Middle East
Number of measures	40		10		17	–
Number of WTO members issuing measures	11		5		11	–

Table 3.2 WTO SPS Committee, 1995–2007 (WTO 2008): Number of phytosanitary trade concerns according to the region of the WTO member that expressed or supported the trade concern

	Phytosanitary trade concerns expressed/supported by WTO members in					
	Developed member countries	Other member countries: Africa	Asia	Europe	Latin America	Middle East
Number of concerns expressed	43	–	6	1	21	1
Number of concerns supported	38	3	20	–	29	–

Table 3.3 WTO SPS Committee, 1995–2007 (WTO 2008): Number of WTO members expressing phytosanitary trade concerns in the WTO SPS Committee in the period 1995 to 2007

	Number of WTO members expressing/supporting concerns in					
	Developed member countries	Other member countries: Africa	Asia	Europe	Latin America	Middle East
Number of WTO members expressing concerns	7	–	4	1	8	1
Number of WTO members supporting concerns	8	2	8	–	12	–

concluded that trade concerns are mostly expressed against the measures of major importing countries, and that trade concerns are practically all expressed by major exporting countries. Phytosanitary measures of LDCs have not been the subject of a trade concern and LDCs have not raised plant health trade concerns and have seldom been supporting the trade concerns of others. It seems likely that in particular LDCs do not have the resources to make Pest Risk Analyses, and, accordingly, adapt their phytosanitary measures, and do not have the resources to challenge the Pest Risk Analyses of importing countries.

Subjects of trade concerns included: general concerns on regulations, undue delays on Pest Risk Analysis and the recognition of pest free areas, phytosanitary measures on potato (brown rot, virus), apples (fireblight), wheat and rice (*Tilletia* spp), fruitflies, citrus canker, and growing media. One case concerned phytosanitary issues in relation to Living Modified Organisms. The implementation of the international standard for wood packing material (IPPC ISPM 15) resulted in a substantial number of general concerns, in particular the introduction date of national measures based on the Standard as industry needed to invest to meet the requirements. Full dispute settlement cases concerned phytosanitary measures relating to codling moth and fireblight of apples. The number of phytosanitary trade concerns has fallen substantially in the last two years.

In comparison, Food Safety concerns dealing with pesticide residues in food are not often raised (10 cases), but when raised, they are more wide reaching, addressing the principles of pesticide registration, like maximum residue levels (MRLs) for pesticides not registered in the importing country and MRLs lower than the Codex Alimentarius standards. Issues concerning mycotoxins were raised only a few times (aflotoxins twice, ochratoxin three times). The latter trade concerns were also raised by LDCs.

There were substantial discussions in the SPS Committee on undue delay, in particular in relation to the recognition of pest free areas. Countries may spend very substantial sums to make, for example, areas free of fruit flies (or in the case of zoonoses, free of animal diseases). A number of countries felt that the recognition of pest freedom, and based on this, withdrawal of phytosanitary measures by certain trading partners, was too slow.

Recently there have also been substantial discussions on private standards (WTO 2007b) in the SPS Committee. A number of countries indicated that these standards are more demanding than official SPS standards and that Governments have a responsibility not to make these standards an additional trade barrier.

3.7 Issues and Challenges

A number of studies have been made on the influence of International Standards on market access of producers in developing countries (Jaffee and Henson 2004; Mold 2005; UNCTAD 2007; Worldbank 2005). These studies address chiefly food safety issues and take private standards into account; plant health is much less considered.

It is recognized that third world farmers of large scale farms can adapt to SPS measures of importing countries based on international SPS standards and private standards. Opinions are divided on the ability of small farmers of small scale farms to adapt to these measures and, in particular, to private standards. The additional requirements for production and certification can only be met if they are able to establish cooperative arrangements or participate in outgrowers schemes.

In a number of situations Food Safety measures appear to become stricter. For example, new pesticide legislation in Europe and Japan reduces the number of pesticides available and new, lower, MRLs are set for mycotoxins. Also, private standards are becoming more elaborate.

It is recognized that phytosanitary measures are probably the most important technical barriers to trade. Concerning plant health, there is a general opinion that countries are reviewing their legislation and the justification for their phytosanitary measures. However, the latter is a long term process. This review of legislation and justification might also be a reason for the decline in phytosanitary trade concerns in the SPS Committee.

It has been argued that ISPMs are too complicated and result in phytosanitary measures that make trade impossible for LDCs to meet measures of their trade partners based on these standards. This is unlikely to be true. In all countries, measures were often not based on science and therefore were often arbitrary, erred on the side of caution, and were sometimes used to protect own producers. Countries are now gradually basing their measures on science, changing or removing those that cannot be justified. The ISPMs help to base measures on science and thus provide the framework to remove unjustified trade barriers, and make trade possible where it was not possible before, but without providing too low a level of protection for the sake of trade or simplicity.

National plant health is driven by producers and trade. In contrast to food safety and animal health, there is little consumer concern. Interest, awareness and funding would be raised if more attention were given to the phytosanitary protection of biodiversity and natural ecosystems.

In many countries national plant health is under-funded. Particularly in developing countries, services to protect plant health, to justify national phytosanitary measures, to make regular surveys and to meet requirements of trade partners, to challenge measures of other countries and to provide the necessary services to export remain very limited. Where existing, services will first be provided to support exports, at the expense of the protection of their own agriculture, biodiversity and natural ecosystems. Technical assistance to developing countries on plant health systems remains very limited, in many cases the analysis of what needs to be done is made, but funds for follow-up activities cannot be identified. Technical assistance to countries should be country driven, but plant health authorities often fail to make their case to ministries of finance in decisions on technical assistance and do not connect to major policy objectives such as the Millennium Goals.

In view of the limited resources at national level, it needs to be considered how to make best use of all biosecurity infrastructure in a country. A substantial number of countries are concentrating food safety, animal and plant health in single agencies, others seek other ways to make best use of synergies among those services.

Plant health has a great need of expertise to identify pests and to understand their behaviour and epidemiology. This is especially true if countries wish to justify their measures through Pest Risk Analysis. Unfortunately, specialized expertise in these disciplines has been declining over the years. Pest identification may benefit from biotechnological tools, but this will not hold for epidemiology.

There is an important case for regional cooperation. It is unlikely that countries, in particular small and less developed countries, can perform the required Pest Risk Analyses in a timely manner. Added to this are issues such as cross-border pest free areas and the prevention and control of introductions. Regional cooperation is indispensable to addressing these issues using joint human and financial resources and know-how.

It is likely that in the future more emphasis will be given to pest free areas, areas of low pest prevalence and pest free production sites. The system of recognition of pest free areas is extremely slow and these areas need to be recognized again and again in bilateral negotiations. Ideally, there would be an international mechanism to recognize such areas, but this will most likely be some time away. Discussions have been initiated in the IPPC Commission on Phytosanitary Measures as to whether creating such a system is possible. Similarly, there would be a great advantage to an international system of certified laboratories for identification of pests.

Private "supermarket" standards have gained enormously in importance lately, especially in relation to food safety (and many other factors in plant production and handling, see above). Supermarkets are using these standards to protect themselves from reputational risk. Different procurement systems use different standards leading to a bewildering number of different standards. Views on the effect of such standards on small farmers and on market access vary (see above). Discussions are now starting in the SPS Committee on their validity and the additional trade restrictions resulting from these standards.

Climate change will alter ecological conditions, including the distribution and the potential distribution of hosts, parasites and weeds (FAO 2008b; Sutherst 2008). The ICPP, in its fourth assessment report (UNEP 2007), has drawn attention to these changes but quantification of the effect of climate change on host pest interaction is very limited. A case could be made to take climate change into account in Pest Risk Analysis. However, in view of the many uncertainties surrounding the magnitude of Climate Change and the ecology of hosts and pests, such considerations will result in phytosanitary measures that are more trade restrictive. It is doubtful if such additional trade restrictions would meet the criterion of being based on science when challenged.

References

Bruinsma J (ed) (2003) World agriculture: towards 2015/2030. An FAO perspective. Earthscan Publications, London, 432 pp

Convention on Biological Diversity (2002) Report of the Sixth Meeting of the Conference of the Parties to the Convention on Biological Diversity. UNEP/CBD/COP/6/20. CBD Secretariat Montreal, 342 pp

FAO (1997) International plant protection convention (New revised text as approved by the FAO Conference at its 29th session –November 1997). FAO, Rome, 16 pp. https://www.ippc.int
FAO (2007a) FAO biosecurity toolkit. FAO, Rome,127 pp
FAO (2007b) The State of Food and Agriculture 2007. Part II. World and Regional Review, a long term perspective. FAO, Rome, 135 pp
FAO (2008a) Third session of the Commission on Phytosanitary Measures. Report. FAO, Rome, 25 pp + 20 Appendices
FAO (2008b) Climate-related transboundary pests and diseases. FAO, Rome, 21 pp
Frederiksen RA, Renfro BL (1977) Global status of maize downy mildew. Annu Rev Phytopathol 15:240–275
IPCC (2007) Contribution of Working Group II to the Fourth Assessment Report of the Intergovernmental Panel on Climate Change, Parry ML, Canziani OF, Palutikof JP, van der Linden PJ and Hanson CE (eds) Climate change 2007: Impacts, adaptation and vulnerability. Cambridge University Press, Cambridge, 976 pp
Jaffee S, Henson S (2004) Standards and agri-food exports from developing countries: Rebalancing the debate. Policy Research Working Paper 3348. World Bank, Washington, DC
Mold A (2005) Non-tariff barriers. Their prevalence and relevance to African countries. Africa Trade Policy Centre, Work in progress No 25
Robinson RA (1976) Plant pathosystems. Springer, Berlin, 184 pp
Secretariat to the Convention on Biological Diversity (2000a) Convention on biological diversity. Text and annexes. CBD Secretariat, Montreal, 34 pp
Secretariat to the Convention on Biological Diversity (2000b) Cartagena protocol on the biosafety to the convention on biological diversity. Text and annexes. CBD Secretariat, Montreal, 30 pp
Sutherst RW (2008) Climate change and vulnerability to introductions by plant and animal pests and diseases. In: FAO (ed) Climate-related transboundary pests and diseases. FAO, Rome, pp 8–21
Tushemereirwe W, Kangire A, Ssekiwoko F, Offord LC, Crozier J, Boa E, Rutherford MA, Smith JJ (2004) First report of *Xanthomonas campestris* pv. *musacearum* on banana in Uganda. Plant Pathology 53:802
Legg JP, Tresh JM (2000) Cassava Mosaic Virus disease in East Africa: a dynamic disease in a changing environment. Virus Research 71:135–149
UNCTAD (2007) Food safety and environmental requirements in export markets – Friend or foe for producers of fruit and vegetables in Asia. United Nations, New York and Geneva, 120 pp
United Nations, Department of Economic and Social Affairs, Population Division. World Population Prospects, 2007: The 2006 Revision. Executive Summary. United Nations, New York, 19 pp
Van der Graaff NA (1999) The International Plant Protection Convention. In: Meester G, Woittiez D, De Zeeuw A (eds) Plants and politics. Wageningen Pers, Wageningen, pp 146–164
Worldbank (2005) Food safety and agricultural health standards. Challenges and opportunities for developing country exports. Worldbank, Washington, DC, 142 pp
Worldbank (2007) World development report 2008. Agriculture for development. Worldbank, Washington, DC, 363 pp
WTO (2007a) The results of the Uruguay Round of multilateral trade negotiations, the legal texts. WTO, Geneva, 550 pp
WTO (2007b) Committee on Sanitary and Phytosanitary Measures. Private standards and the SPS agreement. G/SPS/GEN/746, 9 pp
WTO (2008) Committee on Sanitary and Phytosanitary Measures. Specific Trade concerns. G/SPS/GEN/200/Rev 8

Part II
Global Food Security

Chapter 4
ISPP and the Challenge of Food Security

Peter Scott and Richard Strange

4.1 Seventh ICPP Edinburgh 1998: ISPP Challenged to Act

The International Society for Plant Pathology (ISPP), a Society with limited resources, has chosen to address a challenge of enormous proportions – the challenge of global food security. Given our limitations, and with our preference for actions over words, we have asked ourselves what, realistically, can we *do* to make a difference?

At the seventh International Congress of Plant Pathology (ICPP) in Edinburgh in 1998, the seminal topic was Global Food Security: The Role for Plant Pathology. In a special public meeting with this title (James 1998), the organizers set out:

- The enormity of the problem: more than 800 million people did not have adequate food.
- The contribution to it made by crop pests and pathogens: crop diseases were estimated to reduce production by more than 10%.

It became clear from the presentations that plant pathologists could not afford to ignore the juxtaposition of these figures.

At the 1998 Congress, stimulated by a memorable presentation by Dr. Norman Borlaug (Fig. 4.1), winner of the Nobel Peace Prize for his work on food security, ISPP was challenged to take *action* to promote food security. The Society responded by establishing a Task Force on Global Food Security (ISPP 1998). This chapter is mainly about what the Task Force has done since then.

P. Scott (✉)
CAB International, Wallingford, OX10 8DE, UK
e-mail: p.scott@cabi.org

R. Strange
School of Biological and Chemical Sciences, Birkbeck College, University of London, Malet Street, London, WC1E 7HX, UK
e-mail: r.strange@sbc.bbk.ac.uk

R.N. Strange and M.L. Gullino (eds.), *The Role of Plant Pathology in Food Safety and Food Security*, Plant Pathology in 21st Century 3,
DOI 10.1007/978-1-4020-8932-9_4, © Springer Science+Business Media, B.V. 2010

Fig. 4.1 Dr. Norman Borlaug, who challenged ISPP at its Edinburgh Congress in 1998 to act on food security (Photo of 2003, courtesy of United States Department of Agriculture) http://www.fas.usda.gov/icd/stconf/photos/enlargements/borlaugnl062403.jpg

4.2 Poverty and Hunger: Changes Between 1998 and 2008

But first, let us see what has happened to 'the enormity of the problem' of poverty and hunger in the 10 years since the Edinburgh Congress in 1998. Has progress been made?

The UN has set out eight millennium development goals to be attained by 2015 (UN 2008). Millennium Development Goal 1 is to halve, between 1990 and 2015, the percentage of people whose income is less than $1 a day. Trends are in the right direction (Fig. 4.2), especially in South Asia, including India, and even more in East Asia, including China. So there is some basis for encouragement. But there is little comfort for those whose income remains anywhere near $1 per day. So there remains an enormous challenge. Can we be content that more than 30% of people in South Asia and more than 40% in Sub-Saharan Africa subsist on less than $1 a day?

And if food is short now, and the planet has limited agricultural area, what are the implications of the increase in world population? According to the US Census Bureau, world population has actually grown by 13% since the 1998 Congress (Fig. 4.3) and is expected to grow by 38% from 1990 to 2015, the time span for which the millennium development goal has targeted a halving of poverty and hunger (US Census Bureau 2008).

4.3 ISPP's Task Force on Global Food Security: Activities Under Five Headings

Following the 1998 Congress, ISPP's newly convened Task Force on Global Food Security, consisting of plant pathologists from diverse backgrounds, held a Workshop in Bangkok in 1999. The Task Force decided to focus on what ISPP can do to

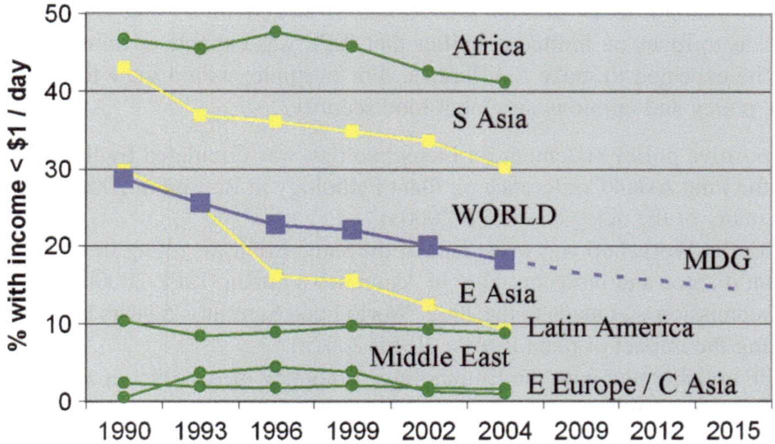

Fig. 4.2 Progress towards Millennium Development Goal (MDG) 1: 'Halve, between 1990 and 2015 the percentage of people whose income is less than $1 a day' (Chen and Ravallion 2007)

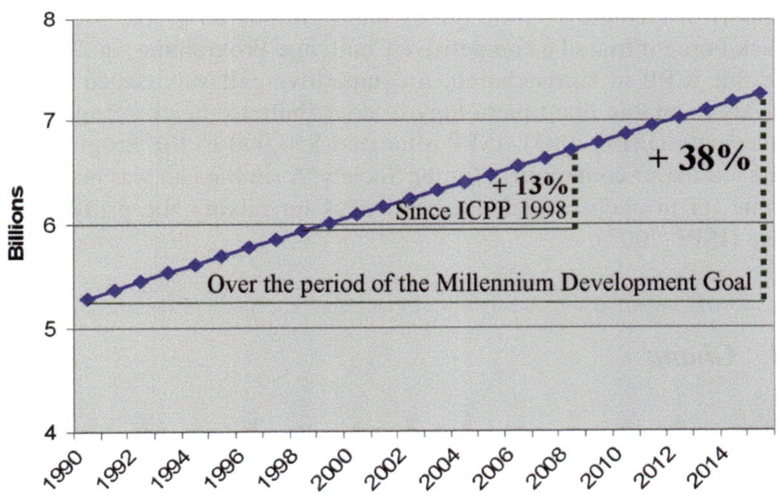

Fig. 4.3 World population. Projections of the US Census Bureau (2008)

deliver tangible results with limited resources (ISPP 1999). Five headings were identified, described as activities:

1. Change public policy and opinions on global food security.
2. Enhance postgraduate training in plant pathology for scientists from a developing country.
3. Quantify the economic impact of some major diseases.
4. Initiate a pilot project for farmer training in simple disease management.
5. Develop the ISPP Web site.

Here, we examine a few examples, especially from activities 1 and 4. The key principle was to focus on limited activities that ISPP was capable of doing and which could be expected to make a difference. For example, with a view to influencing public policy and opinions on global food security:

- A positive policy statement on biotechnology was circulated by the organizers of the First Asian Conference on Plant Pathology in Beijing in 2000, and signed by many of the delegates (ISPP 2000a).
- A Media Workshop was organized at the same event, resulting in reports about plant disease and biotechnology in Asian news media (ISPP 2000b).
- A submission was made to the 2002 'World Food Summit – 5 years later', emphasizing the impact of plant disease (ISPP 2002a).
- ISPP collaborated with the Entomological Society of America in a presentation and a debate on biosafety issues in 2002 (ISPP 2002b).

4.4 Challenge Programme

In support of changing public policy and opinions on global food security, the Task Force initiated a competitive Challenge Programme. In 2003, linked to the 8th ICPP in Christchurch, a competitive call was issued for project proposals to enable plant pathology to contribute to the challenge of global food security (ISPP 2003). ISPP allocated $50,000 to the Programme over 3 years – a major commitment for the Society. A similar call was made in 2007, this time for projects with a specific focus on raising the profile of plant disease (ISPP 2007).

4.4.1 Ghana

The successful project answering the 2003 call was from the Crops Research Institute, Kumasi, Ghana, led by plant pathologist Emmanuel Moses. He attended the 9th ICPP in Torino at the invitation of ISPP. He described the project at the first Keynote Session of the Congress (Moses 2009).

The purpose of the project was to develop, within 3 years, appropriate strategies to control cassava diseases in Ghana, with an emphasis on training farmers to understand that disease limits their production, and then to recognize and manage the diseases that affected their crops.

The project focused on what could be done in practical terms to help farmers to understand how they can grow better crops by demonstrating good practices in the field and by talking to them and their families in group sessions (Fig. 4.4).

The Reports of the project (2008a) contain a number of striking quotations, such as these:

Fig. 4.4 Farmers and their families at a Farmer Field School in Avemedra, Ghana, sharing experiences on cassava diseases with Dr. Emmanuel Moses

- Famine rarely occurs where cassava is grown.
- Most farmers in Ghana have little or no awareness of plant diseases.
- Farmers wanted healthy planting materials, so Community Multiplication Fields were established.

The third one is set in the context of appreciation of ISPP's contribution to local farming, and looks to the prospect of an ongoing legacy of the project.

The project has also delivered an illustrated booklet, *Guide to Identification and Control of Cassava Diseases* (Moses et al. 2007), with descriptions of all the major diseases of cassava that reduce yield in most parts of Sub-Saharan Africa.

ISPP is proud to see these results of its Challenge Programme initiative, and is delighted that Emmanuel Moses could share them at the 2008 Congress.

4.4.2 South Africa

The Task Force has good expectations of the successful project in its second, highly competitive, award in the Challenge Programme. This was made to Lise Korsten of the University of Pretoria (ISPP 2007). The project is a close match to the wording of the Call for Projects that will 'raise the profile of plant disease, as a contribution to the challenge of global food security'. The project proposal does just that, through the establishment of an Information Hub to create awareness of the importance of plant diseases in food security and in the use of a second hand truck as a mobile demonstration laboratory.

4.5 Postgraduate Training for Plant Pathologists from Developing Countries

The Task Force originally identified Ph.D. training as a specific area for study. This was taken up by Richard Strange of Birkbeck College, University of London, who conducted a questionnaire survey of Ph.D. and MSc training. One of its objectives was to determine how to reverse the decline of postgraduate training in plant pathology.

The results (Strange 2003) confirm that declining funds in developed countries underlie the declining availability of training – especially for students from developing countries. It is suggested that a condition of an award should be that a student from a developing country should work in his or her home country for at least 3 years after training in a developed country.

4.6 The Impact of Plant Diseases

The Task Force identified recognition of the impact of disease as a key element in official and public understanding of why plant disease matters. As a contribution to this, the Task Force supported a review in *Annual Review of Phytopathology* of 'Plant disease: a threat to global food security' (Strange and Scott 2005).

There are sections on:

- What are the threats?
- How serious are the threats?
- How can the threats be minimized?
- The challenge of the future.

It is concluded that catastrophic plant diseases significantly exacerbate the current deficit of food supply. At the political level, the need is for plant diseases to be acknowledged as a threat to food supplies and, in consequence, that there is a need for adequate resources to be devoted to their control.

4.7 A New Journal on *Food Security*

The most recent initiative of ISPP's Task Force is the launch of a new journal – *Food Security: The Science, Sociology and Economics of Food Production and Access to Food*. This concept emerged over a period of 2 years during which support was expressed by a distinguished international group of scientists, sociologists and economists who came to form an Advisory Board.

They were united in holding 'a deep concern for the challenge of global food security, together with a vision of the power of shared knowledge as a means of addressing that challenge'. The concept that emerged from much discussion was of a journal that goes far beyond a single discipline such as plant pathology, far beyond plants, and indeed far beyond agriculture and its adjacent sciences (ISPP 2008).

The headings under which topics are to be grouped are these:

- Global food needs: the mismatch between population and the ability to provide adequate nutrition
- Global food potential
- Natural constraints to satisfying global food needs
- Nutrition, food quality and food safety
- Socio-political factors that impinge on the ability to satisfy global food needs

Natural constraints to satisfying global food needs covers biotic constraints, including pathogens and pests, alongside climate, desertification and flooding, natural disasters, and soil constraints. Socio-political factors include agricultural and food policy, international relations and trade, access to food, financial policy, wars and ethnic unrest.

The journal is published quarterly from March 2009, online and in print. It is an official journal of ISPP, with Richard Strange as Editor-in-Chief. Publication has been made possible through a partnership with the publisher Springer, who share ownership with ISPP. Springer are making the financial investment, matched by ISPP's intellectual investment (Springer 2008).

ISPP has been eager to ensure that the journal is affordably available in developing countries, and this is being achieved through FAO's AGORA scheme, WHO's HINARI, and UNEP's OARE. The first issue, and selected content from subsequent issues, are freely available online. ISPP members, and members of ISPP Associated Societies have privileged access.

The Task Force started in 1998 with Dr. Norman Borlaug – plant pathologist and Nobel Peace Prize Laureate – and he also occupies a key place in its activities in 2008. He has commended this latest initiative of ISPP in a piece prepared for the first issue of the journal (Borlaug 2009), including these quotations:

- 'I commend plant pathologists for taking a lead, through practical actions, in the battle for food security.'
- 'In the 10 years since the Edinburgh Congress, the challenge of global food security has sharpened greatly. I said in 2005 that we will have to double the world food supply by 2050.'
- 'It is therefore particularly timely that ISPP and Springer are launching, as a joint venture, this journal with its topical title and with the breadth of coverage indicated by its subtitle.'

4.8 The Stark Challenge

We end with some stark facts about ISPP and the challenge of food security:

- Nine hundred and sixty-three million people are undernourished (FAO 2008).
- Eighteen thousand children under five die of malnutrition every day (WFP: Morris 2006).

Yet in 1974 the World Food Conference declared: 'Every man, woman and child has the inalienable right to be free from hunger and malnutrition.' (UN 1975).

We have set out some initiatives that ISPP has taken over 10 years, through the Task Force on Global Food Security, to address – in a small way – this enormous challenge.

A more important question is this: from now on, what more can ISPP realistically *do* to make some contribution to reaching the World Food Conference's ambitious goal? We hope that the Forum for which this chapter was prepared, and the discussions arising from it, will help to guide us.

References

Borlaug N (2009) Foreword. *Food Security* 1:1

US Census Bureau (2008) International Data Base. Total midyear population for the world. U.S. Census Bureau, Population Division, 4600 Silver Hill Road, Washington DC, USA. http://www.census.gov/ipc/www/idb/worldpop.html

Chen S, Ravallion M (2007) Absolute poverty measures for the developing world, 1981–2004. World Bank Policy Research Working Paper 4211, April 2007

FAO (2008) Number of hungry people rises to 963 million. Newsroom, Food and Agriculture Organization of the United Nations. http://www.fao.org/news/story/en/item/8836/icode/

ISPP (1998) ISPP Task Force on Global Food Security. http://www.isppweb.org/foodsecurity_background.asp

ISPP (1999) ISPP Task Force on Global Food Security. Report on meeting in Bangkok, 13–16 September 1999. http://www.isppweb.org/Society_Reports/rept0099.doc

ISPP (2000a) Executive Committee of the International Society for Plant Pathology Supports Biotechnology. ISPP Policy Statement: First Asian Conference on Plant Pathology, Beijing, August 2000. http://www.isppweb.org/foodsecurity_mins1023.asp#act1.1

ISPP (2000b) ISPP Media Workshop at the First Asian Congress on Plant Pathology, Beijing, August 2000. http://www.isppweb.org/foodsecurity_MediaWorkShop.asp

ISPP (2002a) International Society for Plant Pathology (ISPP), Submission to the World Food Summit: Five Years Later, 10–13 June 2002, Rome. http://www.isppweb.org/foodsecurity_mins1023.asp#act1.3

ISPP (2002b) ISPP and Entomological Society of America: Events on Biosafety, Fort Lauderdale, November 2002. http://www.isppweb.org/foodsecurity_mins1023.asp#act1.4

ISPP (2003) International Society for Plant Pathology Task Force on Global Food Security: Congress Challenge to mark the 8th International Congress of Plant Pathology. http://www.isppweb.org/foodsecurity_congresschallenge.asp

ISPP (2007) International Society for Plant Pathology Task Force on Global Food Security: Challenge Project 2007. http://www.isppweb.org/about_challenge_project_2007_result.asp

ISPP (2008) New ISPP Journal. Food Security: The Science, Sociology and Economics of Food Production and Access to Food. http://www.isppweb.org/about_fs_flyer.asp#4
James C (1998) Global food security. Abstracts, 7th International Congress of Plant Pathology, Edinburgh,UK, August. No. 4.1GF. http://www.bspp.org.uk/icpp98/4/1GF.html
Morris JT (2006) Statement by James T. Morris, Executive Director, World Food Program, to the United States Senate Foreign Relations Committee Hearing 'Ending Child Hunger and Undernutrition Initiative' September 26, 2006, Washington, D.C. http://documents.wfp.org/stellent/groups/public/documents/newsroom/wfp106205.pdf
Moses E (2008) Development of Appropriate Strategies to Control Cassava Diseases in Ghana. Year One Report: 2004–2005. Year Two Report: 2005–2006. Year Three Report: 2006–2007. http://www.isppweb.org/foodsecurity_congresschallangeexecsummary.asp
Moses E (2009) Development of appropriate strategies to control cassava diseases in Ghana. Plant pathology in the 21st century 3:11–24.
Moses E, Asafu-Agyei JN, Adubofuor K, Adusei A (2007) Guide to Identification and Control of Cassava Diseases. Kumasi, Ghana. Crops Research Institute, 32 pp. http://www.isppweb.org/fstf/guide%20to%20identification%20and%20control%20of%20cassava%20diseases.doc
Springer (2008) Food Security: The Science, Sociology and Economics of Food Production and Access to Food. http://www.springer.com/life+sci/agriculture/journal/12571
Strange R (2003) Provision of Ph.D. and M.Sc. Courses in Plant Pathology for Students Normally Domiciled in Developing Countries. http://www.isppweb.org/foodsecurity_phdtraining.asp
Strange RN, Scott PR (2005) Plant disease: a threat to global food security. Annu Rev Phytopathol 43:83–116
UN (1975) Report of the World Food Conference, Rome, 5–16 November 1974. New York, United Nations, 10 pp
UN 2008. The Millennium Development Goals Report 2008. Rome, United Nations Department of Economic and Social Affairs (DESA), ISBN 978-92-1-101173-9, 54 pp

Chapter 5
Globalisation and the Threat to Biosecurity

Harry C. Evans and James M. Waller

5.1 Introduction

During the initial ravages of potato blight in Europe in the 1840s, the hypothesis was advanced that a fungus, parasitic on the foliage, was the cause and not the consequence of the disease (Berkeley 1846). This controversial fungal hypothesis competed with the more-favoured spontaneous generation theory and anticipated Pasteur's germ theory of disease by several decades. However, pre-dating even Berkeley's revolutionary ideas were the pioneering studies in Italy in the 1830s by Agostino Bassi. By careful experimentation, he proved that a fungus was responsible for the notorious, white muscardine disease of silkworms (Major 1944; Steinhaus 1956) and 'so became the first to elucidate the etiology of a contagious microbial disease and to effect the experimental infection of one living organism by another' (Ainsworth 1956). That Berkeley was influenced by Bassi's work, in reaching the conclusion that "the *Botrytis [Phytophthora]* is the immediate cause of the destruction" of the potato crop, is evident from his reference to it and the first ever illustration of the pathogen, *Botrytis [Beauveria] bassiana*, in his seminal publication on the potato murrain (Berkeley 1846).

Both of these host-pathogen associations epitomise the early period of globalisation – removing constraints to movement of goods across geographical and political barriers – as sea routes were opened up for trade and the interchange of new food plants and commodities between Europe and other continents became more frequent: silkworms from Asia during the 12th–13th centuries; potatoes from the New World in the sixteenth century. In both instances, hundreds more years elapsed before their natural enemies caught up with them with disastrous, far-reaching socio-economic and even political repercussions – especially for potato blight – which still reverberate to the present day. The history of this infamous disease occupied centre stage in one of the classic and still relevant accounts of the impact of plant pathogens on global food security (Large 1940).

H.C. Evans and J.M. Waller
CAB International, UK-European Centre, Egham, Surrey, TW20 9TY, UK
e-mail: h.evans@cabi.org

R.N. Strange and M.L. Gullino (eds.), *The Role of Plant Pathology in Food Safety and Food Security*, Plant Pathology in 21st Century 3,
DOI 10.1007/978-1-4020-8932-9_5, © Springer Science+Business Media, B.V. 2010

Here we compare and contrast these past experiences with the modern concepts of globalisation and biosecurity, drawing heavily on the old, as well as the new literature (Large 1940; Anderson et al. 2004).

5.2 The Subject Area Defined

The world has moved on apace since the foundations of plant pathology were laid during the latter half of the nineteenth century, not least in the language employed. Such is the case in point for globalisation and biosecurity; both coined within recent times.

Globalisation appears to have entered the vocabulary within the past two decades, although it has proven difficult to pin down the exact origins, as well as a standard definition. Some argue that it cannot be defined, others offer an array of often contradictory interpretations and, indeed, there is even a specialist publication on this subject in which 114 definitions are detailed (GCSP 2006). Simply put, it is 'to make world-wide in scope or application'. We favour this simple interpretation, paraphrased in the introduction, which automatically implies increased trade and the creation of free markets following deregulation, increased travel and improved communications. With this, of course, comes the increased risk of anthropogenic drivers that facilitate the intercontinental movement of actual or potential pest (*sensu lato*) organisms. Data available from FAO (FAOSTAT@fao.org) show that international trade in agriculture increased by 22% from 2000 to 2005 and recent data from IATA (2008) indicate that passenger travel increased by 21% from 2000 to 2007. There is also a substantial movement of 'food aid' most of which is not subject to normal plant health controls.

Biosecurity involves the policies and measures taken to protect from the damage that these organisms could cause if introduced, either accidentally or deliberately, into new ecosystems. In the present context, therefore, this covers the prevention or mitigation of plant disease. Essentially, we follow the earlier FAO definition – 'Management of all biological and environmental risks associated with food and agriculture' – rather than the later, more encompassing one of the US Government – 'Protection from biological harm of the economy, environment and health of living things from diseases, pests and bioterrorism' (GCSP 2006).

5.2.1 The Threats

The very benefits that globalisation has brought have also created vulnerabilities. Crops producing so well in new lands are open to threats on two fronts.

Firstly, there are the so-called 'new-encounter' pathogens that are not coevolved with the crop species but, nevertheless, can infect the newcomer. Some of these may have coevolved on related native species or they may be relatively minor pathogens on native species but cause damage to the exotic crop because of the particular ecological conditions under which it is grown, for example soil, climate, or absence of antagonists on the introduced crop. These 'new-encounter' pathogens also present a threat to the crop species in its native habitat should they ever arrive there.

Secondly, there is the risk that the crop's coevolved pathogens may catch up with their host in its new environment to cause major problems, as illustrated by the classic examples quoted in the introduction. Table 5.1 lists some other coevolved pathogens that have spread and caught up with their crop hosts, causing significant problems. The damage to the crop may be especially severe, either because of the ecological conditions referred to above or because the crop has become adapted or selected to the new environment and lost some of the resistance that was present in the original crop genotypes in their native environment. These factors have been exacerbated by globalisation. As market preferences have become more dominant, so particular crop varieties have tended to be grown more widely and there has been a general reduction in crop diversity, thus increasing vulnerability to epiphytotics.

Introduced pathogens may also interact with those already present and there are several pertinent examples of new virulent hybrid pathogens developing as a consequence of such encounters such as has occurred with Phytophthora alni, P infestans and Ophiostoma novo-u*lmi* (Brasier and Duncan 1999; Brasier 2001; Schardl and Craven 2003).

Globalisation can also bring more insidious threats. Economic imperatives and easy bulk transport have meant that crop production has become more specialised or concentrated in certain areas and many countries now rely more heavily on imported food. Plant disease problems can easily disrupt this flow of food as witnessed by the recent problem with Karnal bunt on wheat where a number of imported cargoes have been rejected due the detection of the fungus in the consignment and where import restrictions have been imposed. (Sansford et al. 2008).

Globalisation has also widened the threat of terrorism, partly through the ready availability of information on the internet, and bioterrorism raises the distinct possibility that plant pathogens can be used to attack a nation's major crops (Wheelis et al. 2002).

Table 5.1 Recent intercontinental spread of some pathogens of major crops

Pathogen	Crop Disease	Origin	Spread
Puccinia arachidis	Groundnut rust	S. America	Global
Phakopsora pachyrhizi	Soybean rust	Asia	Global
Mycosphaerella fijiensis	Black Sigatoka, Black leaf streak	Pacific Islands	Global
Tilletia indica	Karnal bunt of wheat	Asia	Mexico, USA, S. Africa
Xanthomonas axonopodis pv. citri	Citrus bacterial canker	Asia	S. America then USA
Phytophthora infestans A2 mating type	Potato late blight	Mexico	Global
Hemileia vastatrix	Coffee rust	Africa	Global

Buddenhagen (1977) reviewed the origins of epiphytotics on tropical crops and concluded that they arise from an intensified union of crop and pathogen following some type of separation. Three broad categories could be identified:

1. Those that result from intensification or re-encounter with long-term coevolved pathogens
2. Those that arise from the pathogen's ability to overcome a genetic selection created through plant breeding
3. Those caused by new-encounter pathogens resulting either from movement of a crop or a pathogen to a new region or continent

Even more recent and novel terminology – for many plant pathologists, at least – is currently being used within the context of globalisation and biosecurity to define better the increasing threat from invasive parasitic organisms. This derives from the medical and veterinary fields and has now been applied to plant pathogens in a pivotal paper (Anderson et al. 2004). Clearly, these definitions and the concepts behind them need to be brought to a wider audience of plant pathologists. ICPP 2008 offers the ideal launch pad and the Global Food Security theme is the perfect vehicle. We make no excuses for sequestering the ideas and conclusions of Anderson et al. (2004) to fuel the paper.

Emerging infectious diseases (EIDs) denote those pathogens that have: increased in geographical or host range, or in incidence; changed pathogenesis or newly evolved; recently been discovered or recognised (Lederberg et al. 1992; Daszak et al. 2000).

Pathogen pollution describes the anthropogenic movement of disease-causing organisms outside their natural geographic or host species range. This may involve 'catch-up' with exported coevolved hosts as well as the development of 'new encounter' diseases on new hosts (e.g. *Phytophthora cinnamomi* on the native flora of Australia). Equally, new encounter diseases on exotic hosts represent a pathogen pollution threat to hosts in their native areas (see below under 'Cocoa') (Daszak et al. 2000; Cunningham et al. 2003).

5.3 Drivers of EIDs

Anderson et al. (2004) analysed the major taxonomic groups of plant pathogens causing EIDs, using data from a global electronic reporting system which contains details of plant disease epiphytotics, including the known or posited factors driving the outbreaks. In order of predominance, the causal agents were found to be viruses (47%), fungi (30%) and bacteria (16%). The most important drivers for each group are now summarised.

5.3.1 Pathogen Pollution

Invariably, the principal driver was identified as introduction – accounting for 71% of the outbreaks for viruses, 40% for fungi, and 56% for bacteria. The mechanisms involved are varied: traded goods, agricultural and horticultural, imported officially through regulatory channels; germplasm for field trials or experiments; unofficial or unregulated movement of plants, especially garden species; deliberate introduction of selected pathogens (agri- or bioterrorism). Through recent rapid expansion of globalisation, the volume of imported food shipments into the USA has increased five-fold since the 1990s (Polyak 2004), whilst world trade in agricultural and horticultural commodities is on a similar steep exponential curve; now estimated at over U$850 billion p.a. (Heather and Hallman 2008). International trade in seeds alone may account for up to 10% of this total, and thus may be a potent driver since the incidence of seed-borne pathogens can be high (McGee 1977).

Anderson et al. (2004) concluded that EIDs can directly be linked to this 'increasing volume of globalised trade': a factor previously and repeatedly identified by plant pathologists over the years, from Large (1940) through Klinkowski (1970), Thurston (1973), Kingsolver et al. (1983) and Wilson (1987), to Van Alfen (2001). This has been highlighted more recently by animal disease epidemiologists due to "the unprecedented speed and volume of international travel and trade" which offers "more opportunities than ever before for pathogens to be carried, unwittingly or otherwise, to new ecological horizons" (Cunningham et al. 2003).

The pathogens involved may either be coevolved – catching up with their hosts over time, often protracted in the case of coffee rust – or new encounters, in which endemic plants or crops might be especially vulnerable to alien, new-encounter pathogens (Buddenhagen 1977; Waller 1984; Anderson and May 1986; Ploetz 2007a). Indeed, practitioners of biological control for the management of invasive alien species (IAS), most notably weeds and arthropod pests, have always operated on the premise that coevolved natural enemies are invariably left behind when organisms are moved from their centres of origin or diversity and, therefore, that this increased fitness, or freedom from natural-enemy pressure, constituted one of the principal drivers of invasiveness. Thus, the deliberate introduction of coevolved natural enemies is a potent weapon in the armoury deployed against IAS, and has been exploited successfully for over a century using arthropod agents (parasitoids, predators, herbivores).

However, for staple food crops, repeated introductions to improve the genetic base has been the norm: inevitably, therefore, their coevolved pathogens would eventually catch up, even if it took several centuries, as in the case of potato and grape diseases in Europe (Large 1940). Furthermore, there is the risk that new-encounter pathogens may also be distributed. As also shown for potato blight, introductions of germplasm have been continuous, leading to the global exchange of virulent and fungicide-resistant strains (Fry and Goodwin 1997), as well as mating types (Tantius et al. 1986).

A recent detailed analysis of new EIDs in the UK over a 35-year period revealed that, of the 2234 new records, 45 were considered capable of causing economic or environmental damage (Jones and Baker 2007). The majority were on horticultural produce, reflecting the increasing international trade and diversity in this category and perhaps the inability of current inspection and legislation to safeguard against the risk.

5.3.2 Weather

This was identified as one of the main drivers of bacterial (44%) and fungal (41%) EIDs, but not of viruses (5%), and was associated with unusual weather events, especially higher rainfall or uneven rainfall patterns (Anderson et al. 2004). The phenomenon of climate change is now being increasingly accepted and linked directly to globalisation. As highlighted recently, this could be especially critical in the tropics – where food security is invariably problematic – because there may be greater climatic variability, adding extra uncertainty to decision making in plant-disease management (Garrett et al. 2006).

As well as impacting on and driving epiphytotics in well-established plant diseases, increased precipitation could also affect previously unrecorded or pre-existing minor pathogens and, thereby, may result in the emergence of 'new' major diseases; whilst, also favouring the establishment and/or upsurge of invasive alien pathogens (Rosenzweig et al. 2001; Harvell et al. 2002), such a scenario is thought to have occurred for maize disease caused by the grey leaf blight fungus, *Cercospora zeae-maydis*, in the USA (Anderson et al. 2004).

The incidence and severity of coffee leaf rust, *Hemileia vastatrix*, is limited by altitude. The disease is seldom troublesome at more than 1,500 m above sea level in the equatorial regions of the Andes and the East African highlands, as temperatures, especially at night, are too cool (Waller et al. 2007), but global warming could remove this constraint. Predictions in Brazil suggest that only minor increases in temperature through global warming events could increase significantly the pressures on coffee production from leaf rust (Delgado et al. 2004). There is experimental support for this from a study of pathogens of meadow plants in the USA, which revealed an increase in plant damage with warmer temperatures (Roy et al. 2004), presumably because of the increased amount of time for growth and reproduction, as well as better inter-seasonal survival. Other impacts of global climate change on plant diseases have been reviewed by Chakraborty et al. (2008).

5.3.3 Vectors

Globalisation also increases the risks of movement of insects, specifically the vectors of virus diseases. In addition to carrying new viruses, later-arriving vectors may also catch up with their coevolved, previously introduced viruses which may have gone undetected due to the absence of specialised vectors – or remained of minor incidence due to the inefficiency of local vectors – and, thus, emerge as new diseases or suddenly be associated with major outbreaks. Changes in vector populations have been attributed to 16% of virus EIDs (Anderson et al. 2004). Diseases caused by geminiviruses have been particularly linked with the global movement of their whitefly vectors (Morales and Anderson 2001).

5.3.4 Evolution–Adaptation

The sudden appearance of more virulent pathogens can be the result of increased pathogen pollution when novel strains and new species arise due to intra- and inter-species recombination of the new arrivals with compatible mating types of local or naturalised species. Hybridisation can thus be a potent driver of EIDs, conferring on the pathogen enhanced ability to overcome host resistance in the coevolutionary arms race. Anthropogenic driven, inter-specific hybridisation has now been documented for a number of EIDs, especially *Phytophthora* diseases of tree hosts (Brasier and Duncan 1999; Brasier 2001; Schardl and Craven 2003; Ersek and Nagy 2008).

Globalisation has also driven changes in agricultural practices following the demand for more and varied crop products. Expansion into new areas, often agro-ecologically marginal as urbanisation expands, intensified crop production methods, and the use of new, 'standardised' genotypes can all be considered aspects of globalisation that have increased vulnerability to plant diseases. This may be by: increased selection pressure for the emergence and spread of new virulence (e.g. southern corn leaf blight); increased predisposition to damage by changing agriculture practices (loss of rotations, fallows etc.); or simply, by the increasing importance of plant disease losses as the investment and value of crops increase (Waller 1984).

5.4 Case Studies of EIDs

5.4.1 Food Crops

Four global, staple food crops have been recognised: maize, potato, rice, and wheat (Harlan 1995). However, we also include a fifth, cassava, which is an important subsistence crop in many tropical regions, especially in sub-Saharan Africa. The main EIDs, defined as posing a major constraint (actual or potential) to global production, are discussed briefly in relation to these selected crops.

5.4.1.1 Potato

We return to the blight pathogen of Berkeley (1846) – or, for more impact, the Irish potato famine fungus (Goodwin et al. 1994) – only to find that it is not actually a fungus but belongs to the kingdom *Stramenopila*, more related to plants than the animal antecedents of the kingdom *Fungi*. What would he have thought? What would certainly have surprised him is the fact that we are still at the mercy of the pathogen. This continuing resurgence, especially during the 1990s, is attributed to more aggressive and fungicide-resistant strains arriving from north-western Mexico and replacing resident or naturalised strains (Goodwin et al. 1994, 1995, Fry and Goodwin 1997). Even more alarming is that the devastating outbreaks of the 1840s were probably caused by a single clone of the A1 mating type of the pathogen: now that both mating types are widespread, due to continuous anthropogenic movement fuelled by globalisation, new recombinations and rapid evolution of pathotypes could make the disease even more intractable (Goodwin et al. 1995). Moreover, the production of oospores now enables the pathogen to overwinter in the soil and in the absence of living host material (Drenth et al. 1995).

The international trade in ware potatoes and the ease with which the plant propagates has provided a ready route for the spread of potato pathogens. The import of potatoes from Mexico to Europe following poor harvests due to drought were the most likely source of the A2 mating type of *P. infestans* (Fry and Goodwin 1997). Also, the establishment of *Ralstonia solanacearum* in solanaceous plants along the River Thames in UK is considered to have arisen from effluent contaminated with waste from imported potatoes that were infected (Elphinstone et al. 1998).

The mechanics of pathogen pollution by *P. infestans* were not addressed by Berkeley (1846), although he did allude to, and recognised that blight outbreaks had occurred earlier in North America. Large (1940) speculated that importation of exotic herbarium material was one possible channel of introduction and, somewhat tongue-in-cheek, commented on the many thousands of specimens received by Berkeley from overseas. However, he considered that the most probable entry was either via botanical gardens – 'the field experiments which were being made in every country on the acclimatisation of foreign economic crops' – or trade routes – 'Then in the first glorious years of Queen Victoria's reign, there was a great increase in mercantile traffic' (Large 1940). This was the burgeoning period of globalisation as the British Empire flexed its muscles and transported planting material around the world with no thought to the pests and pathogens it may have harboured. Ironically, potato blight actually contributed to increased globalisation, and hence more risk of pathogen pollution, since its severe socio-economic impact in Ireland led to the repeal of the Corn Laws, allowing for free trade, especially between Europe and North America.

5.4.1.2 Wheat

In the 1990s, Karnal bunt caused by *Tilletia indica* was posing a threat to wheat production – particularly in the major exporting countries of the USA and Australia (Bonde et al. 1997; Murray and Brennan 1998) – not because of direct disease damage, but because of the detrimental impact on seed quality by even minor contamination and the subsequent knock-on effect on world trade. Its sudden and unexpected emergence in Mexico from its restricted natural range in northern Asia in the early 1970s, can only have been via movement of germplasm, especially since it was found in experimental plots. Indeed, there have been suggestions that Karnal bunt was then carried in germplasm to most wheat-growing regions of the world from such field trials over the following decade (Babadoost 2000).

This source was recognised subsequently and access to experimental plots in Mexico was subject to restriction. However, the disease remains absent from Europe, Africa (except S. Africa) and S. America (except Brazil) but remains a plant health risk to these areas. There have been several instances where grain shipments have been rejected due to the presence of the disease and Sansford et al. (2008) provide a detailed assessment of the threat that the disease poses to the European Plant Protection Organisation area.

Whether or not such a route, albeit more indirect, can be implicated in the appearance of a new race of black stem rust (Ug99) – carrying a unique combination of virulence to all known resistance genes of wheat – in Uganda in the late 1990s, is open to debate. However, it was first detected in an experiment station trialling a wide range of new wheat varieties (Stokstad 2007). Because of the favourable year-round conditions of wheat cultivation in the highlands of East Africa, allowing for a build-up of pathogen populations, this region is a known hotspot for rust evolution (Saari and Prescott 1985). The winds flowing around the Intertropical Convergence Zone and moving along the Rift Valley, have now dispersed the rust as far north as the Yemen, threatening the Asian bread basket and the national economies, as well as the livelihoods of millions of small farmers (Singh et al. 2006, Stokstad 2007).

5.4.1.3 Cassava

This crop is an important staple food throughout sub-Saharan Africa and, over recent decades, has been the subject of several high-profile pest invasions from the South American centre of origin, impacting on the food security of the many millions who depend upon it. The spectre of long-term famine has been averted by adopting the classical biological control approach, and alien pest outbreaks have been managed successfully following the introduction of coevolved natural enemies. For example, over recent decades Africa has witnessed the accidental introduction of a number of exotic arthropod pests which have threatened both food security and livelihoods. Cassava mealybug arrived from South America to catch up with its coevolved host with dramatic actual and potential impacts on a vital subsistence crop. Surveys in the Neotropics revealed a natural-enemy complex associated with cassava mealybugs: subsequent selection, screening, mass production and release of a parasitoid resulted in spectacular control of the pest (Neuenschwander 1996). Follow-up projects against cassava mite and mango mealybug (ex India) enjoyed similar success that, critically, has been sustained (Alene et al. 2006). Perhaps there is a lesson here for a similar innovative approach to the management of invasive alien pathogens.

In contrast, however, the geminiviruses causing cassava mosaic diseases appear to have originated in Africa as new-encounter pathogens. A highly virulent form of cassava mosaic disease (CMV) was reported from Uganda in the late 1980s and was apparently associated with dual infections of *African cassava mosaic virus* (ACMV), occurring west of the Rift Valley, and *East African cassava mosaic virus* (EACMV), found east of the Rift Valley A recombinant variant (EACMV-UG2) was also detected in disease plants (Zhou et al. 1997) and also shown to be responsible for the severe form of the disease (Pita et al., 2001; Sseruwagi et al. 2004). The disease appears to have spread from the border area with Sudan and the Democratic Republic of Congo, where it developed undetected in this region of insecurity and caused successive crop failures as it migrated southwards throughout East and Central Africa. Disaster was averted only by concerted international efforts to identify, multiply and release resistant material (Legg and Thresh 2000). The indications are that movement of germplasm or planting material into East Africa brought in the exotic ACMV (Anderson et al. 2004). The inter-relationships between the viruses, their recombinants and vectors continue to evolve (Sseruwagi et al. 2004).

5.4.2 Commodity or Cash Crops

Sometimes termed secondary food crops, commodities are included here because of the socio-economic importance in developing countries, particularly those small-farmer crops that generate not only local income but also foreign exchange.

5.4.2.1 Coffee

The major constraint to production worldwide is coffee leaf rust, the history of which is as intriguing as that of potato blight, especially since the Rev. Berkeley was also involved in the first taxonomic description of the causal agent, *Hemileia vastatrix* (Large 1940; McCook 2006). All evidence points to the fact that movement from its East African origin into Asia in the nineteenth century was anthropogenic – and repeated many times – as coffee selections were transported between the colonies to improve quality and yield in order to supply the demands of Europe in an increasingly globalised world. This was particularly true for India, and it should come as no surprise, therefore, that under the high selection pressure provided by the early deployment of hybrid cultivars, pathogen variation has been correspondingly high and by far the largest number of pathotypes has been recorded from India (Gouveia et al. 2005; Waller et al. 2007). However, the means of arrival of the rust in the Neotropics, over a century later, is more contentious. Theoretically, the urediniospores could have been carried across the Atlantic from Africa in high-altitude air currents (Bowden et al. 1971). However, anthropogenic movement seems more probable and the subsequent jump-spread across the Amazon basin and to Central America was almost certainly by anthropogenic means (Waller 1979).

Coffee berry disease is currently restricted to Africa, but remains a significant threat to other coffee-producing countries. The disease arose when the tetraploid *Coffea arabica,* which evolved in Ethiopia, was reunited with its ancestral diploid progenitors in the Great Lakes region of East Africa. The causal agent, *Colletotrichum kahawae*, first emerged when *Coffea arabica* was planted in Kenya close to the

Ugandan border and was probably selected from the highly variable populations of *Glomerella cingulata* and its *Colletotrichum* anamorphs that occur on and within the tissues of wild *Coffea* species and to which it is closely related (Bridge et al. 2008).

5.4.2.2 Cocoa

This crop exemplifies the phenomenon of new-encounter pathogens. Throughout its exotic palaeotropic range, novel pests and diseases have emerged from the obscurity of forest hosts to become major constraints to production. Amongst the new-encounter pathogens are *Cocoa swollen shoot virus* (CSSV) and *Phytophthora* spp. in West Africa, and the vascular streak die-back fungus (*Oncobasidium theobromae*) in Asia (Thorold 1975; Ploetz 2007a, b). Unusually, the two main South American pathogens have not reached the Old World, almost certainly because the initial germplasm introductions were restricted in number and area of origin (Lower Amazon), which led to genetic uniformity, whilst later ones from more diverse Upper Amazon sources were directed through third and even fourth-quarantine countries and often controlled by national organisations or the chocolate companies themselves (Bartley 2005). Is this a lesson for international centres which supply germplasm directly to producing countries?

Nevertheless, it is the two South American pathogens which give greater cause for concern because of their relatively narrow distribution – although both are now on an invasive front having recently crossed geographic barriers (Evans 2002) – and their perceived greater impact on tree health and/or yield (Evans 2007, Ploetz 2007b). Both are now placed in the genus *Moniliophthora*, belonging to the overwhelmingly saprophytic, basidiomycete order Agaricales (Aime and Phillips-Mora 2005). *Moniliophthora perniciosa* is the causal agent of the notorious witches' broom disease which led to the demise of cocoa cultivation in several South American countries in the late nineteenth and early twentieth centuries, as it spread, almost certainly with human assistance, from its Amazonian centre of origin. Frosty pod rot, caused by *M. roreri*, was originally confined to the north-western side of the Andes but, since anthropogenically breaching the Andean barrier as the eastern Amazonian region of Ecuador was colonised in the 1970s following the discovery of petroleum, it has now dispersed unchecked along the eastern Cordillera to all the cocoa-growing areas of Peru. It poses a potent threat to Bolivia and, especially, Brazil – the most important producer in the Americas (Evans 2002, 2007).

It is considered worthwhile analysing in more detail the history of cocoa and witches' broom disease because it truly encapsulates the title of this paper. Cocoa has the distinction of being the only major crop not cultivated in its centre of origin – the Upper Amazon region – whilst becoming globalised at an early stage in the world history of farming. Its value was appreciated by the Mesoamericans at least three millennia ago (Henderson et al. 2007). So much so, that the product itself became the luxury food of the elite and the standard currency in ancient Mexico (Thorold 1975, Henderson et al. 2007). Literally, the seeds of globalisation – as exemplified by trade, capitalism and power – were sown with this crop.

Globalisation continued apace after the discovery of the New World when cocoa became the fashionable beverage in sophisticated European circles during the seventeenth–eighteenth centuries. Later, during the 1800s, after chocolate was developed as a luxury food, it made fortunes for plantation owners in the Americas as international demand and trade increased. However, the threat of witches' broom disease loomed large as the pathogen emerged or was accidentally carried from its wild forest hosts to devastate plantations in north-eastern South America towards the end of the nineteenth century. This prompted the movement of cocoa cultivation to new regions far removed from the Amazon basin, especially central Brazil (Bahia State) and West Africa, forming the backbone of their economies. The latter region, and indeed the Palaeotropics as a whole, remains free of witches' broom disease, where cocoa provides much needed foreign exchange and some degree of political stability.

Cocoa was the chosen crop for development of the Brazilian Amazon in the 1970s, a project that would make Brazil the world leader in cocoa exports, predicted at over 1 million tons. Underpinning this ambitious plan were the much-trumpeted, high-yielding Upper Amazon varieties, selected for resistance to the *M. perniciosa* strain(s) which destroyed the cocoa industry in Trinidad. The plant breeders' optimism was short-lived as endemic, virulent races emerged from wild cocoa or other *Theobroma* spp in the forest and the expansion phase quickly changed to containment in order to guard against the pathogen reaching the disease-free plantations of Bahia. A cordon sanitaire was enforced, with strict quarantine measures, backed by an awareness campaign. In the late 1980s, however, the disease was reported from Bahia and, despite concerted attempts at eradication, it spread rapidly through the region. Annual production plummeted from over 400,000 tons to less than 150,000 tons and many of the cocoa estates folded. Since it had been posited that the fragile basidiospores could not have reached Bahia naturally, having to pass the formidable barrier of the Caatinga, an immense arid zone of thorn forest in north-east Brazil, it was assumed that the introduction was accidental via cocoa germplasm or its commercially significant relatives such as, *Theobroma grandiflorum* (Evans 2002).

Only recently, however, has the murky truth been revealed in what is probably the first authenticated example (non-governmental, at least) of bioterrorism (Evans 2007). The story involved a plan by socialist opposition supporters (Petistas) to destabilise the power of the pro-government, right-wing cocoa barons by introducing witches' broom disease. Despite attempts to retract the article, the 'confession' is highly credible: matching reports of the sighting of brooms artificially suspended in the cocoa canopy at the outbreak locations; whilst confirming that the site in south-west Amazon – where the diseased material was said to have been collected – is also the source of the exotic pathotype in Bahia, according to molecular evidence. The perpetrators could not have envisaged that such an amateur attempt at bioterrorism would be so spectacularly 'successful'; the government power base was broken in Bahia, with far-reaching socio-economic, as well as ecological impacts (Evans 2002, 2007).

5.5 Tackling EIDs

5.5.1 *Plant Health*

The long-term solution to tackling EIDs is through an effective global phytosanitary system and it is germane to include here the somewhat wistful plea of Large (1940): 'If the nations could only resolve their political and economic differences just sufficiently to be able to do that – to set up an uncorrupted and worldwide organisation to fight the real common enemies of Mankind; the destructive insects and the pathogenic fungi, bacteria and viruses – then there would be a new page to turn over in the history of civilisation'. That this statement was made in the early days of the Second World War only adds to its poignancy and demonstrates the optimism and political naivety of a true scientist. The immediate post-war creation of the UN, and with it the FAO, has gone some way to fulfilling these aspiration, especially the more recent obligation for countries to report plant disease outbreaks to the International Plant Protection Convention (FAO 1995). However, the onus is still on individual countries and so can easily founder on those that do not have adequate diagnostic infrastructures or those that may choose not to disclose EIDs because of economic reasons – often directly linked to globalisation.

Threats posed by the entry of exotic organisms have long been recognised and national/regional lists of quarantine pests are now based on pest risk assessment (PRA) procedures. CABI's Invasive Species list (under the Global Invasive Species Programme) also includes plant pathogens considered to be invasive. However, these lists can only include known organisms about which there is sufficient information to undertake a PRA. Potential threatening organisms might not be listed:

- Because of unclear taxonomic identity or because they are sub-specific taxa (e.g. races or mating types)
- Because they are not economic pests in their native habitat (e.g. through natural constraints, such as biocontrol mechanisms)
- Because they have changed from a minor to a major pest status through genetic adaptation, through agricultural change (e.g. new crop cultivars, new cultivation practices), or through climatic change that favours epidemiological competence (e.g. increased inter-seasonal survival, reduced temperature limitations)

The efficacy of plant-health regulations is also dependent on sampling and detection procedures that can never provide a complete safeguard. The above limitations, coupled with the sheer volume of the movement of commodities and people associated with globalisation, suggests that the threats to biosecurity posed by the movement of plant pathogens seem likely to increase.

The UK government's 'Foresight' programme has examined the likely future requirements needed to tackle new threats from infectious diseases to humans, animals, and plants. Within this programme, systems that would be required for use against such future threats were evaluated by a project studying the detection and identification of infectious diseases (Anon 2007). It recommends the development of:

- Novel information technology systems for early detection of disease events
- Lab-based systems for characterisation of new or newly drug- or chemicaly-resistant/ virulent pathogens
- Hand-held and portable devices for detection and characterisation of infectious diseases
- High through-put screening systems for disease detection at airports and sea ports

We suggest that some lessons could be learned from the two commodity crops that are most threatened by EIDs – cocoa and rubber. Whilst the arrival of the South American fungal pathogens of cocoa in the Palaeotropics could spell economic disaster for individual countries and millions of resource-poor farmers, the impact of rubber leaf blight would be even more dramatic with far-reaching global repercussions, both political and economic (Evans 2002).

How have these coevolved host-pathogen associations been kept apart for the last 100 years or so, despite escalating globalisation?

- The original imported germplasm was limited, ludicrously so for rubber (Evans 2002), and only in recent decades has attention been focussed on improving the genetic base.
- Subsequent importations have always gone through third-country quarantine.
- The responsibility for research and breeding has been with individual countries or industry and never concentrated in an international centre, and, therefore, not subject to multiple turnover and continuous exchange of germplasm.

Considerable work was undertaken from the 1930s on cocoa in Trinidad at the behest of the Colonial Government. This involved searching for and breeding from cocoa material originating from the Upper Amazon centre of diversity with the specific aim of selecting for resistance to witches' broom disease. The Trinidad collection has since been exploited to the full and forms the basis of the high-yielding material which, arguably, has saved the cocoa industry of West Africa, and underpinned that of south-east Asia (Bartley 2005). Crucially, all germplasm has passed through a third-country quarantine for phytosanitary checks, funded in the most part by the chocolate industry itself. Effectively, this has been the quarantine guardian and not disparate countries or organisations. Similarly, Malaysia has ensured that any agricultural material from the Neotropics, related or not to rubber, goes through an intermediate holding and processing system.

International agricultural research institutes, such as the CGIAR centres, charged with gathering and exchanging germplasm have not adopted this third-country quarantine strategy for obvious logistical reasons. However, the fact cannot be ignored that some exotic pest and disease outbreaks have occurred in or around international centres or in-country research stations, and can be linked to movement of crop germplasm (Waller 1984). Seeds, for example, are notoriously difficult to screen for diseases, and many are vectored this way (McGee 1977), so the problems faced are enormous. Even experimental seed lists cannot be guaranteed to be free of extraneous seeds. This was highlighted by research undertaken at IRRI which demonstrated that both incoming and outgoing rice seed shipments were contaminated with weed seeds, amongst the most notorious of these being itch grass, *Rottboellia cochinchinensis* (Huelma et al. 1996). It is not surprising, therefore, that this highly invasive weed has now appeared in a disparate range of countries around the world, especially in the Americas – Bolivia, Colombia, Ecuador, Peru, Mexico, USA (Milhollon and Burner 1993). Moreover, although the species is endemic in Thailand, exotic biotypes resistant to the local coevolved fungal pathogens have recently invaded a research station undertaking international field trials, posing a threat to both agricultural and natural ecosystems (Ellison 1992).

Clearly, improved phytosanitary regulations are needed, especially for germplasm exchange. The challenge will be getting this balance right, so that new plant health standards for both the movement of germplasm and agricultural produce do not overly restrict international crop breeding programmes and global trade.

5.5.2 Resistance

This is both short- and long-term, depending on the crop, and has been instrumental in maintaining global food security. But, as has been shown recently from black stem rust of wheat, and now also for coffee leaf rust, resistance can be lost when pathogen pressure is high – and this is particularly true in multi-varietal field trials. Moreover, as seen with potato blight, resistance is prone to the arrival of exotic, virulent strains of the pathogen (Fry and Goodwin 1997). Growing ‘resistant’ crops in new areas can also lead to disaster, as aggressive pathotypes emerge from wild populations, as happened with the cocoa Scavina hybrids in north-west Brazil (Evans 2002).

Large (1940) ended his treatise on a somewhat pessimistic note when he summarised the conflicting views of plant breeding for disease resistance: ‘Those who believed in genes posited the existence of an eternal quality, R, which they could take from wild plants, build into the genetical constitution of cultivated ones, and so make them disease resistant forever. Those who thought, not in terms of mathematical abstractions, but of the green flux of ever-changing nature, saw little hope of such permanency and no end to man’s labours in defending the crops upon which he depended for life.’

What thoughts for genetically modified crops?

5.5.3 New Approaches

As Anderson et al. (2004) conclude: ‘Reducing the threats and impacts of plant EIDs will require novel approaches to integrated research...’. It is unlikely that the existing technology can achieve the crop production levels to match those of the expanding human populations. Increasing pressures on land use and from invasive alien pests and diseases – as well as climate change linked with drought, demand that promising lines of research need to be pursued. In this context, the potential of fungal endophytes warrants investigation.

There is now compelling evidence that fungal endophytes are important components of plants in natural ecosystems and that these help their hosts to overcome both abiotic and biotic stresses (Rodriguez et al. 2004). On-going work is showing that plant species moved outside of their centres of origin or diversity are impoverished in their coevolved endophytic mycobiota, and, although such endophyte-deficient neotypes have increased fitness in exotic habitats, since there is no trade-off with endophytes and a lack of natural-enemy pressure, they may be highly vulnerable to attack if, and when, their specialist or coevolved natural enemies ever catch up (Evans 2008). This forms the basis of the endophyte-enemy release hypothesis to explain both the invasiveness of alien weed species and the spectacular success of some classical biological control projects in which the release of a single, coevolved natural enemy has resulted in rapid and dramatic population decline of the target weed.

Thus far, few studies have been targeted at the endophytes of the major food crops, especially cereals, and only one has specifically involved the classical biological control approach – searching for and surveying for endophytes in the centre of origin of wheat – with no apparent application of the encouraging results (Marshall et al. 1999). We feel that this promising line of research should be pursued, concentrating on wild populations and related species of the staple food crops: *Oryza* spp. (in north-east India); *Triticum* spp. (in Turkey); *Zea* spp. (in Mexico), *Manihot* spp. (in South America); *Solanum* spp. (in Meso- and South America). Screening endophyte isolates for their ability to enhance the resistance of their plant hosts to pests and diseases, as well as to drought, would be priority criteria. Integrating endophyte-enriched crops with plant breeding programmes for more durable resistance to both abiotic and biotic stresses would be a challenging but rewarding goal.

In conclusion, an integration of different approaches that act independently, but are additive in effect, is advocated for the more effective and sustainable management of EIDs. However, this will require more detailed biological information and a more proactive approach in developing and applying new systems.

References

Aime MC, Phillips-Mora W (2005) The causal agents of witches' broom disease and frosty pod rot of cacao (chocolate, *Theobroma cacao*) form a new lineage of Marasmiaceae. Mycologia 97:1012–1022

Ainsworth GC (1956) Agostino Bassi, 1773–1856. Nature (Lond) 177:255–257

Alene AD, Manyong VM, Coulibaly O (2006) Responding to food supply shocks through global partnerships in technology development and transfer: the case of the IITA-led biological control of cassava mealybug in Sub-Saharan Africa. Outlook Agr 35:255–261

Anderson PK, Cunningham AA, Patel NG, Morales FJ, Epstein PR, Daszak P (2004) Emerging infectious diseases of plants: pathogen pollution, climate change and agrotechnology drivers. Trends Ecol Evol 19:535–544

Anderson RM, May RM (1986) The invasion, persistence and spread of infectious diseases within animal and plant communities. Philos Trans R Soc (Lond) Ser B 314:533–570

Anon (2007) Foresight: Detection and identification of infectious diseases project

One-Year Review April 2006–May 2007

Babadoost M (2000) Comments on the zero-tolerance quarantine of Karnal bunt of wheat. Plant Dis 84:711–712

Bartley BGD (2005) The genetic diversity of cacao. CABI Publishing, Wallingford, Oxon, 341 pp

Berkeley MJ (1846) Observations, botanical and physiological, on the potato murrain. J Hort Soc (Lond) 1:9–34

Bonde MR, Peterson GL, Schaad NW (1997) Karnal bunt of wheat. Plant Dis 81:1370–1377

Bowden J, Gregory PH, Johnson CG (1971) Possible wind transport of coffee leaf rust across the Atlantic Ocean. Nature (Lond) 229:500–501

Brasier CM (2001) Rapid evolution of introduced plant pathogens via interspecific hybridization. BioScience 51:123–133

Brasier CM, Duncan JM (1999) Origin of a new *Phytophthora* pathogen through interspecific hybridization. Proc Natl Acad Sci U S A 96:5878–5883

Bridge PD, Waller JM, Davies D, Buddie AG (2008) Variability of *Colletotrichum kahawae* in relation to other *Colletotrichum* species from tropical perennial crops and the development of diagnostic techniques. J Phytopathol 156:274–280

Buddenhagen IW (1977) Resistance and vulnerability of tropical crops in relation to their evolution and breeding. Ann N Y Acad Sci 287:309–326

Chakraborty S, Luck J, Hollaway G, Freeman A, Norton R, Garrett KA, Percy K, Hopkins A, Davis C, Karnosky DF (2008) Impacts of global change on diseases of agricultural crops and forest trees. CAB Rev 3:54. http://www.cababstractsplus.org/cabreviews

Cunningham AA, Daszak P, Rodriguez JP (2003) Pathogen pollution: defining a parasitological threat to biodiversity conservation. J Parasitol 89:578–583

Daszak PL, Cunningham AA, Hyatt AD (2000) Emerging infectious diseases of wildlife–threats to biodiversity and human health. Science 287:443–449

Delgado AE, Silveira Pinto H, Helminsk AM (2004) Impacto das mudancas climaticas no zoneamento agroclimatico do cafe no Brasil. Pesqu Agropecu Bras 39:1057–1064

Drenth A, Janssen EM, Grovers F (1995) Formation and survival of oospores of *Phytophthora infestans* under natural conditions. Plant Pathol 44:86–94

Ellison CA (1992) Mycoherbicide control of *Rottboellia cochinchinensis*: a viable alternative? Plant Prot Q 7:163–165

Elphinstone JG, Stanford HM, Stead DE (1998) Detection of *Ralstonia solanacearum* in potato tubers, *Solanum dulcamara* and associated irrigation water. In: Prior P, Allen C, Elphinstone J (eds) Bacterial wilt diseases: molecular and ecological aspects. Springer, Berlin

Ersek T, Nagy ZA (2008) Species hybrids of the genus *Phytophthora* with emphasis on the alder pathogen *Phytophthora alni*: a review. Eur J Plant Pathol 122:31–39

Evans HC (2002) Invasive neotropical pathogens of tree crops. In: Watling R, Frankland JC, Ainsworth AM, Isaac W, Robinson CH (eds) Tropical mycology 2: Micromycetes. CABI Publishing, Wallingford, Oxon, pp 83–112

Evans HC (2007) Cacao diseases–the trilogy revisited. Phytopathology 97:1640–1643

Evans HC (2008) The endophyte-enemy release hypothesis: implications for classical biological and plant invasions. In: Julien MH, Sforza R, Bon MC, Evans HC, Thatcher PE, Hinz HL, Rector BG (eds) Proceedings of the twelfth international symposium on biological control of weeds. CABI Publishing, Wallingford, Oxon, pp 20–26

FAO (1995) International standards for phytosanitary measures: Principles of plant quarantine as related to international trade. Rome, Secretariat of the International Plant Protection Convention

Fry WE, Goodwin SB (1997) Resurgence of the Irish potato famine fungus. BioScience 47:363–371

Garrett KA, Dendy SP, Frank EE, Rouse MN, Travers SE (2006) Climate change effects on plant disease: genomes to ecosystems. Annu Rev Phytopathol 44:489–509

GCSP (2006) Definitions of globalization: A comprehensive overview and a proposed definition. Switzerland, Geneva Centre for Security Policy

Goodwin SB, Cohen BA, Deahl KL, Fry WE (1994) Migration from northern Mexico as the probable cause of recent genetic changes in populations of *Phytophthora infestans* in the United States and Canada. Phytopathology 84:553–558

Goodwin SB, Sujkowski LS, Fry WE (1995) Rapid evolution of pathogenicity within clonal lineages of the potato late blight disease fungus. Phytopathology 85:669–676

Gouveia MMC, Ribeiro A, Varzea VMP, Rodrigues CJ (2005) Genetic diversity in *Hemileia vastatrix* based on RAPD markers. Mycologia 97:396–404

Harlan JR (1995) The living fields. Cambridge University Press, UK

Harvell CD, Mitchell CE, Ward JR, Altizer AP, Dobson AP, Astfeld RS, Samuel MD (2002) Climate warming and disease risks for terrestrial and marine biota. Science 296:2158–2162

Heather NW, Hallman GJ (2008) Pest management and phytosanitary trade barriers. CABI Publishing, Wallingford, Oxon

Henderson JS, Joyce RA, Hall GR, Hurst WJ, McGovern PE (2007) Chemical and archeological evidence for the earliest cacao beverages. Proc Natl Acad Sci U S A 104:18937–18940

Huelma CC, Moody K, Mew TW (1996) Weed seeds in rice seed shipments: A case study. Int J Pest Manage 42:147–150

IATA (2008) World air transport statistics, 52nd edn. IATA

Jones DR, Baker RHA (2007) Introduction of non-native plant pathogens into Great Britain, 1970–2004. Plant Pathol 56:891–911

Kingsolver CH, Melching JS, Bromfield KR (1983) The threat of exotic plant pathogens to agriculture in the United States. Plant Dis 67:595–600

Klinkowski M (1970) Catastrophic plant disease. Annu Rev Phytopathol 8:37–60

Large EC (1940) The advance of the fungi. J. Cape, London

Lederberg J, Shope RE, Oaks SC (eds) (1992) Emerging infections: Microbial threats to health in the United States. National Academy Press, Washington, DC

Legg JP, Thresh JM (2000) Cassava mosaic virus disease in East Africa: a dynamic disease in a changing environment. Virus Res 71:135–149

Major RH (1944) Agostino Bassi and the parasite theory of disease. Bull Hist Med 16:97–107

Marshall D, Tunali B, Nelson LR (1999) Occurrence of fungal endophytes in species of wild *Triticum*. Crop Sci 39:1507–1512

McCook S (2006) Global rust belt: *Hemileia vastatrix* and the ecological integration of world coffee production since 1850. J Global Hist 1:177–195

McGee DC (1977) Plant pathogens and the worldwide movement of seeds. APS Press, St. Paul, MN

Milhollon RW, Burner DM (1993) Itchgrass (*Rottboellia cochinchinensis*) biotypes in world populations. Weed Sci 41:379–387

Morales FJ, Anderson PK (2001) The emergence and dissemination of whitefly-transmitted geminiviruses in Latin America. Arch Virol 146:415–441

Murray GM, Brennan JP (1998) The risk to Australia from *Tilletia indica*, the cause of Karnal bunt of wheat. Australas Plant Pathol 27:212–225

Neuenschwander P (1996) Evaluating the efficacy of biological control of three exotic homopteran pests in tropical Africa. Entomophaga 41:405–424

Pita JS, Fondong VN, Sangare A, Otim-Nape GW, Ogwal S, Fauquet CM (2001) Recombination, pseudorecombination and synergism of geminiviruses are determinant keys to the epidemic of severe cassava mosaic disease in Uganda. J Gen Virol 82:655–665

Ploetz RC (2007a) Diseases of tropical perennial crops: challenging problems in diverse environments. Plant Dis 91:644–663

Ploetz RC (2007b) Cacao diseases: important threats to chocolate production worldwide. Phytopathology 97:1634–1639

Polyak MG (2004) The threat of agroterrorism. Georget J Int Aff 2004:31–39

Rodriguez RJ, Redman RS, Henson JM (2004) The role of fungal symbioses in the adaptation of plants to high stress environments. Mitig Adapt Strat Glob Change 9:261–272

Rosenzweig C, Iglesias A, Yang XB, Epstein PR, Chivian E (2001) Climate change and extreme weather events: implications for food production, plant diseases and pests. Global Change Hum Health 2:91–104

Roy BA, Gusewell S, Harte J (2004) Response of plant pathogens and herbivores to a warming experiment. Ecology 85:2570–2581

Saari EE, Prescott JM (1985) World distribution in relation to economic losses. In: Roelfs AP, Bushnell WH (eds) The cereal rusts, vol. 2, distribution, epidemiology and control. Academic, Orlando, FL, pp 259–298

Sansford CE, Baker RHA, Brennan F, Ewert B, Gioli A, Inman A, Kinsella A, Magnus HA, Miglietta F, Murray GM, Port-Puglia A, Porter JA, Rafoss T, Riccioni L, Thorne F (2008) The new Pest Risk Analysis for *Tilletia indica*, the cause of Karnal bunt of wheat, continues to support the quarantine status of the pathogen in Europe. Plant Pathol 57:603–611

Schardl CL, Craven KD (2003) Interspecific hybridization in plant-associated fungi and oomycetes: A review. Mol Ecol 12:2861–2873

Singh RP, Hodson DP, Jin Y, Huerto-Espino J, Kinya MG, Wanyera R, Njau P, Ward RW (2006) Current status, likely migration and strategies to mitigate the threat to wheat production from race Ug99 (TTKS) of stem rust pathogen. CAB Rev: Perspect Agri, Vet Sci, Nutr Nat Res 1:1–13

Sseruwagi P, Rey MEC, Brown JK, Legg JP (2004) The cassava mosaic geminiviruses occurring in Uganda following the 1990s epidemic of severe cassava mosaic disease. Ann Appl Biol 145:113–121

Steinhaus EA (1956) Microbial control, the emergence of an idea. Hilgardia 26:107–157

Stokstad E (2007) Deadly wheat fungus threatens world's breadbaskets. Science 315:1786–1787

Tantius PM, Fyfe AM, Shaw D, Shattock RC (1986) Occurrence of the A2 mating type and self fertile isolates of *Phytophthora infestans* in England and Wales. Plant Pathol 35:578–581

Thorold CA (1975) Diseases of cocoa. Clarendon Press, Oxford

Thurston HD (1973) Threatening plant diseases. Annu Rev Phytopathol 11:27–52

Van Alfen NK (2001) Whither plant pathology? Phytopathology 91:17–19

Waller JM (1979) The recent spread of coffee rust (*Hemileia vastatrix* Berk. & Br) and attempts to control it. In: Ebbels DL, King JE (eds) Plant health. Blackwell, Oxford, pp 275–283

Waller JM (1984) The influence of agricultural development on crop diseases. Trop Pest Manag 30:86–93

Waller JM, Bigger M, Hillocks RJ (2007) Coffee pests, diseases and their management. CABI Publishing, Wallingford, Oxon

Wheelis M, Casagrande R, Madden LV (2002) Biological attack on agriculture: low-tech, high-impact terrorism. BioScience 52:569–576

Wilson CL (1987) Exotic plant pathogens – who's responsible? Plant Dis 71:863

Zhou X, Liu Y, Clavert L, Muniz C, Otim-Nape GW, Robinson DJ, Harrison BD (1997) Evidence that DNA-A of a geminivirus associated with severe cassava mosaic disease in Uganda has arisen by interspecific recombination. J Gen Virol 78:2101–2111

Chapter 6
Genetic Modification (GM) as a New Tool in the Resistance Toolbox

Thomas Hohn and Gabriele Schachermayr

The Indo-Swiss Collaboration in Biotechnology (ISCB) is much concerned with the theme of the conference "Plant Pathology and Global Food Security". The ISCB is a bilateral research programme, jointly funded by the Swiss Agency for Development and Cooperation (SDC) in Berne and by the Department of Biotechnology (DBT) in New Delhi. It promotes research partnerships between Swiss and Indian research institutions in various areas of biotechnology and fosters technology transfer to the private industry. The ISCB programme focuses on agriculture and environment, research areas which are central for the overall development of the rural and urban populations in India. Currently, the ISCB is funding several joint projects related to the sustainable production of pulses and wheat and to sustainable management of natural resources in semi-arid and rain-fed agricultural systems in India.

The crop based approach was adopted because pulses and wheat are important components in the vegetarian diet, which is still prevalent in India. Pulses or grain legumes are rich in basic amino acids and poor in sulphur containing ones, while the reverse is true for graminaceous grains, such as wheat or rice. Together, they provide a balance in essential amino acids. The choice of these two types of crops is based on the needs of people depending mainly on a vegetarian diet. A mixed diet would provide for sufficient protein nutrition, while a diet composed of, for instance, rice only would lead to severe protein deficiency or protein malnutrition, for example "Kwashiorkor". Interestingly, people from different parts of the world, who are dependent on a vegetarian diet, design dishes composed of products derived from both sources, for example "rice and dahl" (rice, lentils), "falafel" (wheat, chickpea), "taco and beans" (maize, beans), "risi bisi" (rice, peas).

Pulse production in India is hampered by a number of factors. Pulses are sensitive to various biotic stresses such as insects, fungi, bacteria and viruses, as well as to abiotic

T. Hohn (✉)
Basel University, Botanical Institute, Basel, Switzerland
e-mail: hohn@fmi.ch

G. Schachermayr
Ecole Polytechnique Fédérale de Lausanne, Indo-Swiss Collaboration in Biotechnology, Switzerland

R.N. Strange and M.L. Gullino (eds.), *The Role of Plant Pathology in Food Safety and Food Security*, Plant Pathology in 21st Century 3,
DOI 10.1007/978-1-4020-8932-9_6, © Springer Science+Business Media, B.V. 2010

stresses such as drought, salinity and cold. The investments in pulse crop improvement through breeding and advancement of production technology are limited. In this regard pulses can almost be considered as neglected crops. For the subsistence farmer, they are therefore "risk crops". Reducing this risk by providing resistant plants would encourage a more widespread cultivation of pulses and ensure a sufficient supply of them.

The ISCB pulse network (Fig. 6.1) concentrates on two very important pulses in India, chickpea (*Cicer arietinum*) and black gram (*Vigna mungo*), and applies gene technology to improve resistance against drought, insect pests and viruses.

Of the six billion people populating the earth today, one billion are over-nourished, with the consequence of severe obesity, while one billion, living primarily in developing countries, are undernourished and have to spend the largest part of their income on food. Rising food prices, as experienced in connection with the recent food crisis, will cut those off from the food chain, leading to misery, political upheaval, promotion of regional conflicts and migration. Therefore, to achieve global food security and affordable food prices, all possible technologies have to be exploited and this includes in the long run the promising and safe technology of genetic modification. Furthermore, in fighting plant pathogenic insects and viruses,

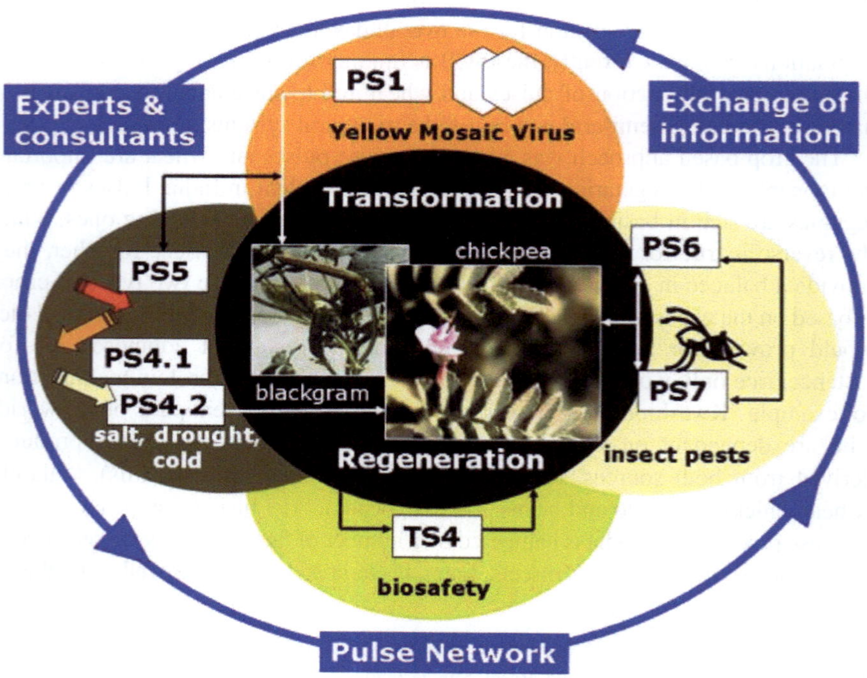

Fig. 6.1 The pulse network of Indo-Swiss collaboration in biotechnology (ISCB) (http://iscb.epfl.ch/). The network combines several research groups with the common goal of working out an efficient transformation system for pulses, which will then be used to engineer insect, virus and abiotic stress resistance (PS1, etc.). The network concentrates on chickpea and black gram. A separate programme (TS4) deals with biosafety aspects

gene technology has a high potential for providing environmentally friendly alternatives to conventional plant protection and for contributing to the reduction of pesticide applications.

In one of the ISCB Pulse Network projects, Professor Sampa Das from the Bose Institute, Calcutta and her co-workers realized that aphids never feed on garlic, but do feed on chickpea. Garlic contains a gene encoding a lectin, which is poisonous for aphids and several other insects, but not for mammals. They isolated the gene and transformed chickpea with it, with the result that aphids avoided the transgenic plant. For humans it would not imply more than consuming a chickpea dish spiced with garlic (published so far as a model system; Dutta et al. 2005).

Karuppanan Veluthambi, Madurai Agricultural University collaborates with our lab at the University of Basel to achieve resistance of black gram against the *Mungbean yellow mosaic geminivirus*. We have chosen "gene silencing" as our strategy: plants recognize double stranded RNA, which accumulates as a byproduct of virus replication (RNA viruses) and transcription (DNA viruses), as foreign. A family of enzymes, named dicer, cleaves this dsRNA into fragments of 21–24 nucleotide lengths. Single-strand versions of these small interfering (si) RNAs are handed over to another enzyme, argonaute, which scans viral and other RNAs for cognate sequences, which it either cleaves or the translation of which it inhibits. Additionally, siRNAs are used to inactivate cognate DNA by methylation and modification of its accompanying histones. Thereby cognate virus sequences are destroyed or at least inhibited. Interestingly, the resulting RNA fragments can be used by a third enzyme, RNA-dependant RNA polymerase to produce more dsRNA, which again becomes cleaved by dicers and again used by argonaute to cleave more cognate RNA, etc., an autocatalytic process. Furthermore, siRNAs or other RNA components of the silencing pathway can be systemically transported throughout the plant, establishing silencing far from the original site of the trigger.

Our strategy is to increase the silencing response by providing a type of transgene that produces dsRNA cognate to the virus sequence. While still working on transformation to obtain transgenic versions of the recalcitrant black gram plants, we applied our constructs transiently to one single leaf each of infected plantlets using a gene-gun. To our surprise, the plantlets recovered from virus infection, producing healthy leaves, but retaining about 0.1% of the virus load of untreated infected plants (Fig. 6.2; Pooggin et al. 2003). Apparently the dsRNA produced from the transfecting plasmid was sufficient to produce siRNAs, which, together with target RNA fragments, was fed into the autocatalytic silencing cycle and transported systemically throughout the plant.

We are presently investigating whether this transient, that is "transgene-free gene technology", which we name "RNA based plant vaccination", can be upgraded for an economically feasible treatment of virus infections in the field. Large scale production of dsRNA pools utilizing fermentation of *Pseudomonas syringae* infected with dsRNA bacteriophage φ6 vectors (Aalto et al. 2007) and application devices developed by Bio-Oz (http://www.bio-oz.co.il/products-bimfield.html) will be tested for this purpose.

Fig. 6.2 Mungbean yellow mosaic virus infected black gram plants. The plant on the right was "vaccinated" by bombardment with a plasmid producing double strand RNA covering the common region of the virus genome

Furthermore, we would like to investigate whether RNA-based plant vaccination can also be used against other plant pathogens, especially nematodes.

Concerning green biotechnology, we envisage that in the near future different stakeholder groups including critically opposed non-governmental organizations ("green NGOs") will work together with plant gene technologists to address the pressing problems of hunger and malnutrition. There is a clear promise that biotechnology can contribute to meeting these challenges and that genetically modified crops will be able to improve agricultural productivity sustainably and enhance local and national food security, while preserving the environment. In this sense we have to be aware of the time pressure: as Mikhaïl Gorbachev said "Those who are too late will be punished by history".

References

Aalto AP, Sarin LP, van Dijk AA, Saarma M, Poranen MM, Arumäe U, Bamford DH (2007) Large-scale production of dsRNA and siRNA pools for RNA interference utilizing bacteriophage φ6 RNA-dependent RNA polymerase. RNA 13:422–429

Chakraborti D, Sarkar A, Mondal HA, Das S (2009) Tissue specific expression of potent insecticidal, Allium sativum leaf agglutinin (ASAL) in important pulse crop, chickpea (*Cicer arietinum L.*) to resist the phloem feeding *Aphis craccivora*. Transgenic Research, 18:529–544

Dutta I, Saha P, Majumder P, Sarkar A, Chakraborti D, Banerjee S, Das S (2005) The efficacy of a novel insecticidal protein, Allium sativum leaf lectin (ASAL), against homopteran insects monitored in transgenic tobacco. Plant Biotechnol J 3:601–611

Pooggin M, Shivaprasad PV, Veluthambi K, Hohn T (2003) RNAi targeting of DNA virus in plants. Nat Biotechnol 21:131–132

Chapter 7
The Role of Plant Pathology and Biotechnology in Food Security in Africa

James M. Onsando and Florence Wambugu

7.1 The Global Food Situation

The world's population is anticipated to increase from 5.7 billion in 1995 to 7.7 billion in 2020 (UN 1996). The highest rates of population growth will be in Asia and Sub-Saharan Africa where 900 million people live on a dollar a day or less (Ahmed et al. 2007). In fact, Sub-Saharan Africa is home to three-quarters of the world's 162 million ultra poor that is those living on less than $0.50 per day (Ahmed et al. 2007). According to the Global Hunger Index, the hot spots of hunger are in Sub-Saharan Africa and Asia (von Grebmer et al. 2008). While economic growth of 2.7% per year is anticipated globally between 1993 and 2020, growth rates are anticipated to be very low in Sub-Saharan Africa. The increase in global food demands between 1993 and 2020 on the other hand will be 41% for cereals, 63% for meats and 40% for roots and tubers. These demands will be higher in Sub-Saharan Africa where currently 23% of the world's hungry and malnourished people live.

7.2 Yield Losses Caused by Diseases and Pests

High crop yields in the developed world have been a product of high yielding germplasm, good soil health and nutrition, good crop husbandry and robust disease and pest management. However, yields in Africa are generally lower. If we consider maize for example, average yields per hectare are 1.8 tonnes for Africa and 5.0 tonnes for the world. Similarly, for bananas the averages are 7.8 tonnes for Africa compared to 18.4 tonnes for the world (Table 7.1). Average yields for other crops in Africa are also less than those for the world (Table 7.1). While many factors may cause these lower yields, diseases and pests do play a significant role (Table 7.2).

J.M. Onsando (✉) and F. Wambugu
Africa Harvest Biotech Foundation International, Whitefield Place, Westlands,
P.O. Box 642-00621, Nairobi, Kenya
e-mail: jonsando@ahbfi.or.ke

R.N. Strange and M.L. Gullino (eds.), *The Role of Plant Pathology in Food Safety and Food Security*, Plant Pathology in 21st Century 3,
DOI 10.1007/978-1-4020-8932-9_7,

Table 7.1 Yields of selected crops: Africa vs. the World (FAOSTAT 2007)

Crop	Africa (Average Yield/ Tonnes/Hectare)	Africa (Yield 00Expressed as % World Yield)	World (Average Yield/ Tonnes/Hectare)
Maize	1.8	36	5.0
Cassava	9.9	81	12.2
Sugarcane	56.8	80	70.9
Sweet potato	4.2	30	13.8
Potato	10.9	66	16.6
Wheat	2.1	75	2.8
Bananas	7.8	42	18.4
Soybean	1.2	52	2.3
Rice	2.5	60	4.2

7.3 Biotechnology and Food Security

The most dramatic changes in agricultural technology in the last quarter of the 20th century have come from biotechnology (Montague et al. 1998). The use of biotechnology for the improved management of pests encompasses disease-free planting material produced through tissue culture and micropropagation, diagnostic techniques such as virus indexing, bio-pesticides such as *Bacillus thurigiensis* (BT) and *Trichoderma* and transgenic crops with increased disease and pest resistance. The use of transgenic technology to manage diseases and pests has attracted considerable interest and activity but the more commercial success stories are in the management of insect pests and weeds. For plant disease management, however, efforts are at the development phase with many cases being at the proof of concept stage, the exception here being the resistance of transgenic papaya to papaya ringspot virus in Hawaii (Ferreira et al. 2002).

7.4 Disease Free Planting Material Obtained Through Tissue Culture

There are several examples where use of tissue culture coupled with virus indexing have given rise to significant levels of disease control with corresponding yield benefits and hence food security. This is particularly true of vegetatively propagated crops such as banana. In Kenya, sourcing clean and virus free banana planting material in adaptable varieties from South Africa (Fig. 7.1) almost doubled banana yields compared with the traditional propagation method of using suckers (Fig. 7.2). Despite the higher initial investment in clean tissue cultured planting material, the increased yields have resulted not only in food security but enhanced incomes as well (Fig. 7.3).

Table 7.2 Losses (%) caused by pathogens, viruses, animal pests and weeds to some crops grown on the African continent (Crop Protection Compendium 2004)

		Losses (%) Caused by:				
Crop	Region	Pathogens	Viruses	Animal Pests	Weeds	Total
Maize	East Africa	13	6	17	19	55
	Southern Africa	10	4	13	15	42
	West Africa	14	6	19	19	58
	North Africa	7	2	9	11	29
Wheat	East Africa	13	3	10	10	36
	Southern Africa	5	2	9	5	20
	West Africa	10	3	10	10	33
	North Africa	9	3	9	9	30
Rice	East Africa	16	2	16	15	49
	Southern Africa	12	2	18	12	44
	West Africa	16	1	18	16	51
	North Africa	5	1	7	7	20
Potatoes	East Africa	23	9	13	11	56
	Southern Africa	10	7	10	9	36
	West Africa	24	9	13	14	60
	North Africa	13	7	10	8	38
Groundnut	East Africa	22	9	16	13	60
	Southern Africa	24	6	17	11	58
	West Africa	22	9	16	13	60
	North Africa	13	6	13	8	40

Fig. 7.1 Certified and clean tissue-cultured banana seedlings

Fig. 7.2 Sucker derived banana on the left and banana grown from tissue culture on the right

7.5 Bio-Pesticides

Biopesticides are preferred in the management of diseases and insect pests in some quarters because of their "green" attributes. These are their target specificity and the fact that they do not leave harmful residues. However, they are not as efficient or as cost effective as chemicals (Montague et al. 1998).

The agents employed as biopesticides include plant products such as neem, insects, fungi, bacteria and actinomycetes. These act by means of antibiosis, predation, competition, parasitism and inhibition of virulence mechanisms of the pest (Strange 2003).

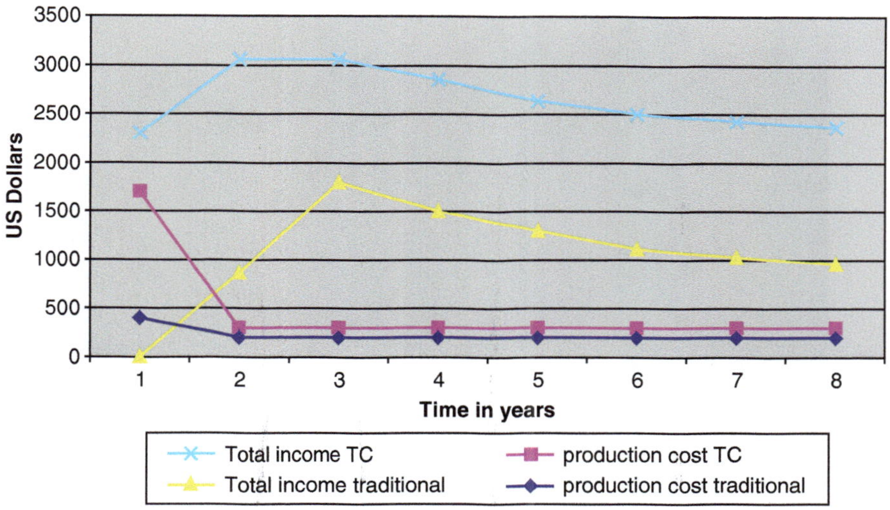

Fig. 7.3 Total production costs and total incomes per hectare from both traditional and tissue culture (TC) bananas on a small scale farm

For the management of plant diseases fungi have been studied particularly in the area of soil borne plant pathogens or root rots. *Trichoderma* tops the list of the most studied fungus in relation to the control of root rots. Its efficacy was first demonstrated in 1930 (Yudelman et al. 1998), and since then it has been widely used and developed globally.

In Africa in vitro and in vivo efficacy of *Trichoderma* has been reported in South Africa against *Sclerotinia sclerotiorum* on soybean (Visser et al. 2008), in Egypt against *Fusarium solani*, *F. oxysporum* and *Macrophomina phaseolina* on grapevines (El-Mohamedy et al. 2008) and in Kenya against *Armillaria* root rot of tea (Onsando and Waudo 1992, 1994). Use of coffee pulp as a substrate in the Kenyan case enhanced the proliferation of *Trichoderma* in the soil and hence bio-efficacy (see Figs. 7.4 and 7.5).

Despite the high efficacy of *Trichoderma*, the full formulation and commercialisation of biopesticides in Africa has not been realised, mainly because of product development costs.

Bacillus thuringienses (*Bt.*) is among the most widely known and researched biocontrol agents. The bacterium produces proteinaceous toxins – *Bt* toxins – that are effective against Lepidoptera (caterpillars), Diptera and Coleoptera. Different strains of the bacterium are used against different insects. The one that is effective against caterpillars has been available commercially for over 30 years in a wettable powder form as a spray against the cotton bollworm. The most significant technological advancement in the use of this biocontrol agent has been the successful engineering of the proteinaceous toxin into the host plant itself for better protection of the entire plant (Monsanto 1997). Bt cotton has been commercialised in South

Fig. 7.4 Showing *Armillaria* infected tea bushes

Fig. 7.5 Showing *Armillaria* healthy infected tea bushes

Africa and Burkina Faso with significant success (Kirsten and Gouse 2002; Vitale et al. 2008). In South Africa despite higher seed cost and additional technology fees, both large scale and small scale farmers realise increased net incomes from *Bt* Cotton due to higher yields and savings on pesticides (see Figs. 7.6 and 7.7). Because of the success story of Bt cotton in South Africa the adoption rose from 0% in 1997 to 80% in year 2001 dropping slightly to 75% in 2002. In Burkina Faso on the yield advantage of Bt cotton was 15% and insecticide sprays were reduced by two-thirds. The profitability gains were in the range of US $ 79 to 154 per hectare (Vitale et al. 2008).

Fig. 7.6 Showing conventional Cotton on the left and Bt Cotton on the right

Fig. 7.7 Showing conventional Maize on the left and Bt Maize on the right

Bt Maize has been successfully deployed in South Africa and Egypt. In South Africa, irrigated and dry land commercial farms surveyed in Mpumalanga, Northern Cape and North West provinces enjoyed statistically significant yield increases of 11% and 10.6% respectively during the 2000 and 2001 seasons (Gouse et al. 2005). In Egypt *Bt* maize was found to be highly resistant to all 3 Egyptian borers with the conventional maize variety showing 78% infestation compounded by opportunistic fungal attacks whereas the Bt Maize was not affected at all (Hamid 2007).

7.6 The Role of Biotechnology in Virus Disease Management

Virus diseases have a significant economic impact on all crops. Studies in several countries have shown that losses due to virus diseases are up to 90% (Muimba-Kankolongo et al. 2008). In the developed world, where production is restricted to relatively few crops on a large scale such as wheat, maize and vegetables, this potential is minimised through the use of insecticides to manage the vectors and through the deployment of virus resistance genes introgressed in costly breeding programmes from related wild species (Maule 2002). In Africa, the challenges are more diverse and less well supported by financial and technical resources and yet this is where the need is greatest.

Biotechnology has a significant role to play in virus disease management. For example, identification of resistance genes effective against different strains of viruses may be facilitated by the employment of a gene gun and marker-assisted breeding (Ariyo et al. 2006). Despite the antipathy in Europe to genetic modification, this technology has and will continue to have an important role to play in food security. The African scenario will, however, remain pathetic unless more resources are put into the development of infrastructure and human skills.

Virus resistance by means of transgenic crops holds great promise. The success story in this area is transgenic papaya in Hawaii, which is resistant to papaya ringspot virus. Here production has increased significantly as a result of the cultivation of the transgenic crop (Ferreira et al. 2002; Gonsalves and Ferreira 2003). In this story, Hawaii's papaya industry faced a potential economic disaster when papaya ringspot virus was discovered in the Puna district of Hawaii Island where 95% of the state's papaya was grown. The disease spread quickly and by 1995 there was a crisis. The introduction of the transgenic virus resistant varieties in 1998 reversed the crisis. Other research in this area has shown great promise but is far from commercialisation.

7.7 Biotechnology in the Management of Fungal Diseases

To-date trials on crops engineered for resistance to fungal pathogens have remained experimental. Technological constraints have contributed to the slower progress in this area.

7.8 Discussion

From the foregoing, it is evident that the application of biotechnology to solve pest and disease problems which affect food security in Africa is limited. This is because most of the effort ends at the experimental and proof of concept phase. The skills and finances to package the efforts for deployment are lacking. This explains why there is a reasonable number of publications in scientific conferences but very little corresponding impact on the ground. However, tissue culture and hence production of clean planting materials for vegetatively propagated crops and development of bio-pesticides have made good progress. Genetic modification for disease control and management on the other hand is predominantly at the development phase. Even in developed countries the commercial successes that have been registered are in the management of insects and weeds. The products in this area which have global application have been achieved through the intervention of the private sector. This brings me to the subject of the ill-preparedness of the public sector to develop products. The scientists in public research institutions globally are very good in the area of technology development up to proof of concept but the skills to carry this through to product development and deployment are lacking. This is a huge bottleneck even in "the North" and, of course, much worse in Africa.

For biotechnology and plant pathology to impact food security positively in Africa, institutional partnerships must be formed whereby the private sector plays a key role in product or technology development while the academic and research institutions develop customised technologies that are efficacious. The third and most important component is funding, capacity building and development of scientific infrastructure. Bilateral funding negotiated by African governments as grants or with the Gates Foundation are the most feasible approaches. This way, African technology efforts will be focused on products that are attractive to African end users and hence exert a market pull that will make the product or technology impactful and sustainable. Without this approach, only the economically and technologically more developed countries such as Argentina, China, India, Indonesia and Brazil will raise the critical amount of funding necessary to support the infrastructure and human resources required to proceed with the techniques that are important for food security in their countries. Thus Africa will remain a continent in which crop productivity will stay at a low level, a situation that is not likely to change as the burden of research and development in developing countries, particularly in Africa, will fall on the public sector (Brenner and Komen 1994) with the inherent bottlenecks already discussed.

The other strategy for Africa is leapfrogging on the technologies or products, rather than starting from scratch. This will involve collaborative efforts with the international private sector and negotiation of intellectual property rights and bio-safety regulations in order to facilitate access to the new processes and technologies.

References

Ariyo AO, Atiri GI, Dixon AGO, Winter S (2006) Virology methods. J Virol Methods 137(1):43–50

Brenner C, Komen J (1994) International initiatives in biotechnology for developing country agriculture: Promises and problems. Technical Paper No. 100. OECD Development Centre, Paris

El-Mohamedy R, Zedan E, Adalla A (2008) Biological soil treatment with *Trichoderma harzianum* to control root rot disease of grapevine (*Vitis vinifera* L.) in newly reclaimed lands in Nobaria province. J Plant Pathol 90(2):132

Ferreira SA, Pitz KY, Manshardt R, Zee R, Fitch F, Gonsalves D (2002) Virus coat protein transgenic papaya provides practical control of papaya ringspot virus in Hawaii. Plant Dis 86:101–105

Gonsalves D, Ferreira S (2003) Transgenic papaya: A case for managing risks of papaya ringspot virus in Hawaii. Online. Plant Health Progress. doi: 10.1094/PHP-2003-1113-03-RV

Gouse M, Pray C, Kirsten JF, Schimmelpfenning D (2005) A GM subsistence crop in Africa: the case of Bt white maize in South Africa. Int J Biotechnol 7(1/2/3):84–94

Hamid IA (2007) Bt. Maize trials in Egypt. Crop Biotech Update

http://www.ifpri.org/media/20071106Deprived/mostdeprivedfindings07.asp. Cited Mar 2009

Kirsten JF Gouse M (2002) Bt. Cotton in South Africa: Adoption and impact on farmer incomes amongst small and large scale farmers. Working Paper 2002-15, University of Pretoria

Maule A (2002) Future prospects for biotechnology in virus disease management. Proceedings of plant pathology and global food security. Imperial College, London, p 19

Monsanto (1997) Monsanto achievements. Plant biotechnology. Monsanto, St. Louis, Missouri, USA

Muimba-Kankolongo A, Apetorgbor M, Begonde DA, Nakobonge G, Wingfield MJ (2008) The many challenges for plant pathology in Africa. J Plant Pathol 90(2):72

Munkvold GP, Hellmich RL (2000) Genetically modified insect resistant maize: Implications for management of ear and stalk diseases. Plant Health Progress. doi: 10.1094/PHP-2000-0912-01-RV

Onsando JM, Waudo SW (1992) Effect of Coffee Pulp on *Trichoderma* spp. in Kenyan Tea Soils. Tropical Pest Management 38(4):376–381

Onsando JM, Waudo SW (1994) Interaction between *Trichoderma* species and *Armillaria* Root Rot fungus of Tea in Kenya. Int J Pest Management 1994(1):69–74

Strange RN (2003) Introduction to plant pathology. Wiley, Chichester, UK, xiii+464 pp

United Nations (UN) (1996) World population prospects: The 1996 revisions. UN, New York

Visser DD, Caldwell PM, McLaren NW (2008) Biological control of *Sclertinia sclerotiorum* using *Trichoderma harzianum* and silcon. J Plant Pathol 90(2):113

Vitale J, Glick H, Greenplate J, Abdennadher M, Traore O (2008) Second-generation Bt. Cotton field trials in Burkina Faso: Analysing the potential benefits to West African farmers. Crop Sci 48:1958–1966

Von Grebmer K, Fritschel H, Nestorova B, Olofinbyi T, Pandya-Lorch R, Yohannes Y (2008) Global Hunger Index. The Challenge of Hunger 2008. IFPRI Issue Brief 54

World Bank (1997) World Bank development indicators. World Bank, Washington DC

Yudelman M, Ratta A, Nygaard D (1998). Pest management and food production. Looking to the future. Food, Agriculture, and the Environment Discussion Paper No 25. International Food Policy Research Institute Washington, DC, pp 1–49

Part III
Mycotoxins

Chapter 8
The Secondary Metabolite Toxin, Sirodesmin PL, and Its Role in Virulence of the Blackleg Fungus

Barbara J. Howlett, Ellen M. Fox, Anton J. Cozijnsen, Angela P. Van de Wouw, and Candace E. Elliott

8.1 Sirodesmin PL, an Epipolythiodioxopiperazine Toxin

Sirodesmin PL is a non-host specific toxin produced by the ascomycete *Leptosphaeria maculans*, which causes blackleg (phoma stem canker) of oilseed rape (*Brassica napus*), the most damaging disease of this crop worldwide (Fitt et al. 2006). As well as causing chlorotic (yellow) lesions on leaves, sirodesmin PL has antibacterial and antiviral properties (Rouxel et al. 1988). Sirodesmin PL is a member of the epi-polythiodioxopiperazine (ETP) class of fungal secondary metabolites, which are characterised by a sulphur-bridged dioxopiperazine ring synthesised from two amino acids (Fig. 8.1a, b). The disulphide bridge confers toxicity by enabling ETPs to cross-link proteins via cysteine residues, and to generate reactive oxygen species through redox cycling.

At least 14 different ETPs are known, and all are produced by ascomycetes (for review see Gardiner et al. 2005). The best-characterised ETP, gliotoxin (Fig. 8.1c), is produced by the opportunistic human pathogen *Aspergillus fumigatus*, as well as *A. terreus*, *A. flavus*, *A .niger*, *Penicillium terlikowskii* and *Trichoderma virens*, a fungus that controls root diseases of plants. Other ETPs with roles in animal diseases include sporidesmins, produced by *Pithomyces chartarum* which infects grasses and causes facial eczema and liver diseases of grazing animals. Chaetomin and chaetocin are produced by *Chaetomium globosum*, a systemic pathogen of immunocompromised humans. The cytotoxicity of some ETPs has made them attractive as potential anticancer agents and accordingly interest in these molecules is increasing.

B.J. Howlett (✉), E.M. Fox, A.J. Cozijnsen, A.P. Wouw, and C.E. Elliott
School of Botany, The University of Melbourne, Vic 3010, Australia
e-mail: bhowlett@unimelb.edu.au

R.N. Strange and M.L. Gullino (eds.), *The Role of Plant Pathology in Food Safety and Food Security*, Plant Pathology in the 21st Century 3,
DOI 10.1007/978-1-4020-8932-9_8, © Springer Science+Business Media B.V. 2010

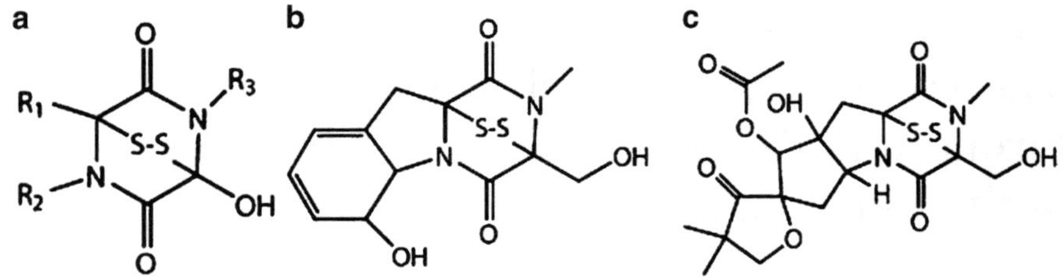

Fig. 8.1 Structures of epithiodioxopiperazines. (**a**) Core epithiodioxopiperazine (ETP) moiety; (**b**) sirodesmin PL; (**c**) gliotoxin

8.2 Biosynthesis of Sirodesmin PL

Sirodesmin PL is derived from serine and tyrosine. Labelling experiments and analysis of putative intermediates have been used to deduce the biosynthetic pathway (Ferezou et al. 1980a, b; Bu'Lock and Clough 1992). A prenyl transferase is predicted to catalyse the addition of a dimethylallyl group to either the dipeptide cyclo-L-tyrosyl-L-serine or free tyrosine, before condensation with serine, to produce the intermediate cyclic dipeptide, phomamide. The genes responsible for the biosynthesis of sirodesmin PL in *L. maculans* were identified 4 years ago by an approach that took advantage of the clustering of genes encoding biosynthetic enzymes for secondary metabolites in fungal genomes (Gardiner et al. 2004). A homologue of prenyl transferase (a dimethyl tryptophan synthetase) from the endophyte *Neotyphodium coenphialum* was identified from an *L. maculans* Expressed Sequence Tag (EST) library. This EST was used to probe an *L. maculans* cosmid DNA library and a cosmid (35 kb) containing the prenyl transferase was sequenced. Regions flanking this prenyl transferase gene (named *sirD*) contained a cluster of 18 genes which, based on best matches to genes in databases, could be assigned roles in sirodesmin PL biosynthesis. These included a two module non-ribosomal peptide synthetase (*sirP*), acetyl transferase (*sirH*), methyl transferases (*sirM* and *sirN*), an ATP-binding cassette (ABC) type transporter (*sirA*), responsible for toxin efflux, and a member of the zinc binuclear cluster ($Zn(II)_2Cys_6$) family (*sirZ*) (Fig. 8.2).

Of the nine cluster genes tested, all were co-regulated and timing of expression was consistent with the production of sirodesmin PL in culture, suggesting the gene cluster is involved in sirodesmin PL biosynthesis. Furthermore, the disruption of the peptide synthetase, *sirP*, resulted in a mutant isolate unable to produce sirodesmin PL, confirming that this gene is essential for the synthesis of this toxin, and that the biosynthetic gene cluster was correctly identified (Fox and Howlett 2008a).

8.3 Regulation of Sirodesmin PL Biosynthesis

Members of the zinc binuclear cluster ($Zn(II)_2Cys_6$) family of transcription factors are unique to fungi (Todd and Andrianopoulos 1997). Genes in this family control production of other secondary metabolites such as aflatoxins, therefore,

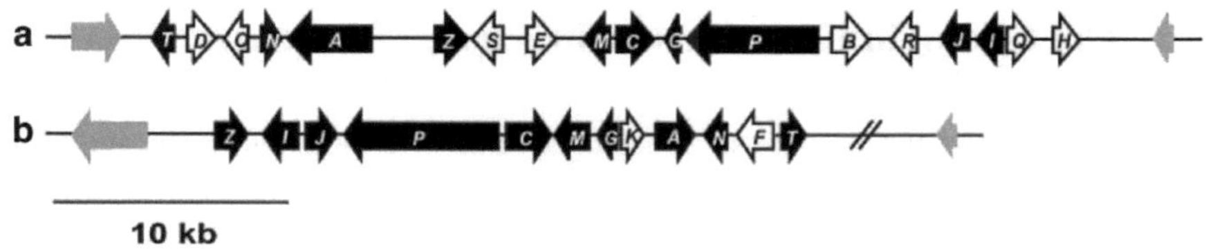

Fig. 8.2 The *Leptosphaeria maculans* sirodesmin PL (**a**) and *Aspergillus fumigatus* gliotoxin (**b**) biosynthetic gene clusters. Common ETP moiety genes (*white text* on *black background*) include those with best matches to non-ribosomal peptide synthetase (*P*), thioredoxin reductase (*T*), methyl transferases (*M* and *N*), glutathione-*S*-transferase (*G*) and cytochrome P450 mono-oxygenase (*C*), amino cyclopropane carboxylate synthase (ACCS) (*I*), dipeptidase (*J*), as well as a transcriptional regulator (*Z*) and a transporter (*A*). Other genes (*black text* on white background) do not have obvious homologues in the other cluster and are thought to be involved in modification of the side chains of the core ETP moiety. These encode cytochrome P450 mono-oxygenases (*F*, *B* and *E*), a prenyl transferase (*D*), an acetyl transferase (*H*), epimerases (*Q*, *S* and *R*), an oxidoreductase (*O*) and a hypothetical protein (*K*) (Gardiner and Howlett 2005). Genes shaded in grey encode proteins with best matches to proteins with no potential roles in ETP biosynthesis. The *forward slash marks* represent a 17 kb region of repetitive DNA (reproduced from Fox and Howlett (2008a), with permission from Elsevier)

SirZ was an obvious candidate for the regulation of sirodesmin PL. RNAi-induced silencing of this gene led to minimal production of sirodesmin PL and very low expression of several of the biosynthetic genes (Fox et al. 2008). Binding sites for $Zn(II)_2Cys_6$ transcription factors consist of conserved terminal trinucleotides, usually in a symmetrical configuration, spaced by an internal variable sequence of defined length (Todd and Andrianopoulos 1997). Such sequences are present in the promoters of most of the sirodesmin PL biosynthetic genes, and it is likely that these are binding sites for the pathway specific transcription factor, sirZ (Fox et al. 2008).

In order to identify regulators of sirodesmin PL biosynthesis upstream of *sirZ*, a library of *L. maculans* insertional mutants has been screened for lack of sirodesmin PL production, using an assay that relies on the antibacterial properties of sirodesmin (E.M. Fox and B.J. Howlett, unpublished, 2009). The insertional mutants were created via *Agrobacterium tumefaciens*-mediated transformation whereby T-DNA is inserted randomly in the genome (Elliott and Howlett 2006). Ten-day-old colonies of individual *L. maculans* mutants growing on an agar plate were assayed for loss of the ability to produce a ring of clearing when overlaid with molten agar containing *Bacillus subtilis*. Five sirodesmin-deficient mutants have been isolated; four of which have insertions in genes with best matches to transcription factors, whilst the other is a hypothetical gene (E.M. Fox and B.J. Howlett, unpublished, 2009). One of the transcription factors was *cpcA*, a general amino acid transcriptional regulator in fungi including *A. fumigatus* (Krappmann et al. 2004). When amino acids are unavailable, fungi activate a complex regulatory network that allows the coordinated expression of a whole suite of genes required for amino acid biosynthesis. The central control element of this network is CpcA. The role of *cpcA* in the regulation of sirodesmin PL biosynthesis is currently being assessed.

8.4 The Gliotoxin Gene Cluster in *Aspergillus fumigatus*

Upon completion of the genome sequence of *A. fumigatus* the gliotoxin biosynthetic gene cluster was identified in this fungus. Ten clustered genes were identified, eight of which were in common with genes in the sirodesmin gene cluster. These genes had an expression pattern consistent with the timing of production of gliotoxin (Gardiner and Howlett 2005). The identity of this cluster was confirmed as the gliotoxin cluster by several research groups via the disruption of the two module non-ribosomal peptide synthetase, *gliP*. The resultant mutant was unable to make gliotoxin (Cramer et al. 2006; Kupfahl et al. 2006), confirming that this indeed was the gliotoxin biosynthetic cluster. Additionally, disruption of *gliZ*, encoding a $Zn(II)_2Cys_6$ transcription factor like sirZ, resulted in a mutant isolate unable to produce gliotoxin (Bok et al. 2006).

Both the *L. maculans* and *A. fumigatus* ETP biosynthetic clusters encode proteins predicted to be involved in the efflux of the toxin from the cell, SirA and GliA respectively. However, while SirA is an ATP-binding cassette (ABC) type transporter, GliA is a Major Facilitator Superfamily transporter (Gardiner et al. 2005).

8.5 Origin and Distribution of ETP Gene Clusters in Ascomycetes

With the increasing number of sequenced genomes becoming available for fungi, the distribution and heredity of gene clusters such as those encoding ETPs can be investigated. Putative ETP gene clusters with most or all of the common eight genes in the sirodesmin and gliotoxin gene clusters have been found in *Magnaporthe grisea*, *Neosartorya fischeri*, *Penicillium lilacinoechinulatum* and *Trichoderma reesei* (Patron et al. 2007). Whether these fungi produce ETPs is unknown. Numerous other ascomycetes, including *Saccharomyces cerevisiae* and *Schizosaccharomyces pombe*, as well as all basidiomycetes that have been sequenced so far lack ETP-like clusters.

Phylogenetic analysis of individual cluster genes has shown ETP gene clusters appear to have a common origin and the cluster has been inherited relatively intact, rather than assembling independently in the different ascomycete lineages (Patron et al. 2007). Although a mechanism describing this complex heredity cannot be conclusively identified, movement of entire clusters by horizontal gene transfer is the most parsimonious explanation of the discontinuous distribution of clusters.

8.6 Role of Sirodesmin PL in Virulence of *Leptosphaeria maculans* on Oilseed Rape

The contribution of sirodesmin PL to virulence of *L. maculans* on oilseed rape (*Brassica napus*) has now been determined unequivocally, due to exploitation of a defined sirodesmin-deficient *L. maculans* mutant, that had a DNA insertion in the non-ribosomal peptide synthetase gene (*sirP*). When the *sirP* mutant was inoculated onto cotyledons of *B. napus*, it caused similar-sized lesions as the wild type isolate, indicating that sirodesmin PL was not a virulence factor at this stage of infection. Subsequently, the mutant caused fewer stem lesions and was half as effective as the wild type in colonising stems, as shown by quantitative PCR analyses (Elliott et al. 2007) (Fig. 8.3). Thus sirodesmin PL contributes to virulence in *B. napus* stems. The expression of two cluster genes, the peptide synthetase, *sirP* and an ABC transporter, *sirA*, was also studied during infection. Fungal isolates containing fusions of the green fluorescent protein gene (GFP) with the promoters of these genes fluoresced 10 days post-inoculation (dpi). This expression pattern was consistent with the distribution of sirodesmin PL in both cotyledons and stems, as revealed by mass spectrometry experiments (Elliott et al. 2007).

It is intriguing to speculate as to why sirodesmin PL contributes to colonisation of stems but not to generation of the necrotic symptoms of lesions in the cotyledons. Sirodesmin PL may suppress plant defences during the biotrophic growth down the stem, or it might play a role in competition between *L. maculans* and other microorganisms in planta, or even on stubble during the saprophytic stage of its life cycle. Germination of several fungal species including *Fusarium graminaerum* and *L. biglobosa* 'brassicae' was inhibited in the presence of a 13 day old colony of the wild type sirodesmin PL-producing isolate when grown in vitro, illustrating the antifungal activity of this molecule (Elliott et al. 2007).

The role that secondary metabolites play in the biology of fungi is elusive. In many cases fungi producing toxins do not rely on growth on a host to complete their life cycle. The most likely advantage of secondary metabolites to organisms that produce such molecules is that they enhance survival. Many such organisms live saprophytically in the soil where they are exposed to a harsh environment with a plethora of competing organisms. Fungal virulence has been proposed to have evolved to protect fungi in such an environment against amoebae, nematodes or other invertebrates that

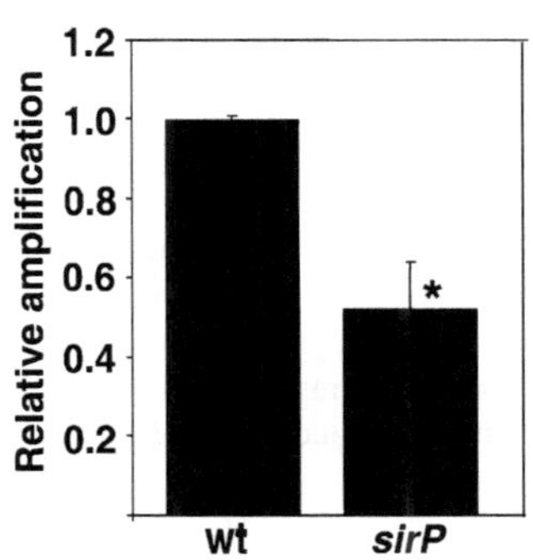

Fig. 8.3 Quantification of fungal biomass in infected stems of *Brassica napus* cv. Monty. Cotyledons and first leaves of *B. napus* cv. Monty were inoculated with wild type (wt) or the sirodesmin-deficient mutant, *sirP* and were analysed at 33 dpi. Genomic DNA was used as a template for quantitative PCR. The relative amount of fungal biomass was calculated as the ratio of the amplification of a fragment of fungal actin to that of plant actin. Each *bar* represents the average of three replicate PCRs on three independent sets of infected stems. The *asterisk* indicates that fungal biomass of wild type is significantly different from that of the mutant

can feed on fungi (Mylonakis et al. 2007); Fox and Howlett (2008b). Secondary metabolite toxins could play a role in such behaviour. The recent availability of defined mutants in the biosynthesis of secondary metabolites will enable this hypothesis to be tested for some molecules and their producing-organisms.

Acknowledgements We thank the Australian Research Council and the Grains Research and Development Corporation for funding our research.

References

Bok JW, Chung D, Balajee SA, Marr KA, Andes D, Nielsen KF, Frisvad JC, Kirby KA, Keller NP (2006) GliZ, a transcriptional regulator of gliotoxin biosynthesis, contributes to *Aspergillus fumigatus* virulence. Infect Immun 74:6761–6768

Bu'Lock JD, Clough LE (1992) Sirodesmin biosynthesis. Austr J Chem 45:39–45

Cramer RA, Gamcsik MP, Brooking RM, Najvar LK, Krikpatrick WR, Patterson TF, Balibar CJ, Graybill JR, Perfect JR, Abraham SN, Steinbach WJ (2006) Disruption of a nonribosomal peptide synthetase in *Aspergillus fumigatus* eliminates gliotoxin production. Eukaryot Cell 5:979–980

Elliott CE, Howlett BJ (2006) Overexpression of a 3-ketoacyl-CoA thiolase in *Leptosphaeria maculans* causes reduced pathogenicity on *Brassica napus*. Mol Plant Microbe Interact 19:588–596

Elliott CE, Gardiner DM, Thomas G, Cozijnsen AJ, Van De Wouw A, Howlett BJ (2007) Production of the toxin sirodesmin PL by *Leptosphaeria maculans* during infection of *Brassica napus*. Mol Plant Pathol 8:791–802

Ferezou JP, Quesneau-Thierry A, Barbier M, Kollmann A, Bousquet JF (1980a) Structure and synthesis of phomamide, a new piperazine-2,5-dione related to the sirodesmins, isolated from the culture medium of *Phoma lingam* Tode. J Chem Soc, Perkin Trans 1: Org Bio-Org Chem (1972–1999), 113–115

Ferezou J-P, Quesneau-Thierry A, Servy C, Zissmann E, Barbier M (1980a) Sirodesmin PL biosynthesis in *Phoma lingam* Tode. J Chem Soc. Perkin Trans 1:1739–1746

Fitt BDL, Brun H, Barbetti MJ, Rimmer SR (2006) World-wide importance of crown canker (*Leptosphaeria maculans* and *L. biglobosa*) on oilseed rape (*Brassica napus*). Eur J Plant Pathol 114:3–15

Fox EM, Howlett BJ (2008a) Biosynthetic gene clusters for epipolythiodioxopiperazines in filamentous fungi. Mycol Res 112:162–169

Fox EM, Howlett BJ (2008b) Secondary metabolism: regulation and role in fungal biology. Curr Opin Microbiol 11:481–487

Fox EM, Gardiner DM, Keller NP, Howlett BJ (2008) A $Zn(II)_2Cys_6$ DNA binding protein regulates the sirodesmin PL biosynthetic gene cluster in *Leptosphaeria maculans*. Fungal Genet Biol 45:671–682

Gardiner DM, Cozijnsen AJ, Wilson L, Pedras MSC, Howlett BJ (2004) The sirodesmin biosynthetic gene cluster of the plant pathogenic fungus *Leptosphaeria maculans*. Mol Microbiol 53:1307–1318

Gardiner DM, Howlett BJ (2005) Bioinformatic and expression analysis of the putative gliotoxin gene cluster of *Aspergillus fumigatus*. FEMS Microbiol Lett 248:241–248

Gardiner DM, Waring P, Howlett BJ (2005) The epipolythiodioxopiperazine (ETP) class of fungal toxins: distribution, mode of action, functions and biosynthesis. Microbiology 151:1021–1032

Krappmann S, Bignell EM, Reichard U, Rogers T, Haynes K, Braus GH (2004) The *Aspergillus fumigatus* transcriptional activator CpcA contributes significantly to the virulence of this fungal pathogen. Mol Microbiol 52:785–799

Kupfahl C, Heinekamp T, Geginat G, Ruppert T, Hartl A, Herbert H, Brakhage AA (2006) Deletion of the *gliP* gene of *Aspergillus fumigatus* results in loss of gliotoxin production but has no effect on virulence of the fungus in a low-dose mouse infection model. Mol Microbiol 62:292–302

Mylonakis E, Casadevall A, Ausubel FM (2007) Exploiting amoeboid and non-vertebrate animal model systems to study the virulence of human pathogenic fungi. PLoS Pathog 3:e101

Patron NJ, Waller RF, Cozijnsen AJ, Straney DC, Gardiner DM, Nierman WC, Howlett BJ (2007) Origin and distribution of epipolythiodioxopiperazine (ETP) gene clusters in filamentous ascomycetes. BMC Evol Biol 7:174

Rouxel T, Chupeau Y, Fritz R, Kollmann A, Bousquet J-F (1988) Biological effects of sirodesmin PL, a phytotoxin produced by *Leptosphaeria maculans*. Plant Sci 57:45–53

Todd RB, Andrianopoulos A (1997) Evolution of a fungal regulatory gene family: the $Zn(II)_2Cys_6$ binuclear cluster DNA binding motif. Fungal Genet Biol 213:388–405

Chapter 9
Biological and Chemical Complexity of *Fusarium proliferatum*

Robert H. Proctor, Anne E. Desjardins, and Antonio Moretti

9.1 Taxonomy and Identification

Fusarium is an ascomycetous fungus that is characterized by production of elongated, multiseptate macroconidia. Fusaria that also produce microconidia in chains and/or false heads have been reported as pathogens on a diverse range of plants for ~100 years (Desjardins 2006; Marasas et al. 1984; Sheldon 1904). For decades these fungi were regarded as one species, *F. moniliforme*. But, as information on host range and morphological variation has accumulated, *F. moniliforme* has been resolved into an increasing number of distinct species. In one taxonomic system, it was resolved into four species, *F. anthophilum*, *F. moniliforme*, *F. proliferatum*, and *F. subglutinans* (Nelson et al. 1983). In subsequent systems, *F. moniliforme* was further resolved into *F. thapsinum* and *F. verticillioides*, while *F. subglutinans* was resolved into *F. circinatum*, *F. guttiforme*, *F. saccharri*, and *F. subglutinans* (Klittich et al. 1997; Leslie and Summerell 2006; Nirenberg and O'Donnell 1998). The resolved species tend to have relatively narrow host ranges. For example, *F. thapsinum* occurs primarily on sorghum and *F. verticillioides* primarily on maize, while *F. circinatum*, *F. guttiforme*, *F. saccharri*, and *F. subglutinans* occur primarily on pine, pineapple, sugarcane, and maize, respectively (Leslie and Summerell 2006). Among the species resolved from *F. moniliforme*, *F. proliferatum* is unusual in that it has an extraordinarily broad host range. It is a pathogen of plants as diverse as asparagus, fig, maize, onion, palm, pine and wheat.

R.H. Proctor (✉) and A.E. Desjardins
US Department of Agriculture*, National Center for Agricultural Utilization Research, Agriculture Research Service, Peoria, IL, USA

A. Moretti
National Research Council, Institute of Sciences of Food Production, Bari, Italy

*Mention of trade names or commercial products in this publication is solely for the prupose of providing specific information and does not imply recommendation of endorsement by the U.S. Department of Agriculture

R.N. Strange and M.L. Gullino (eds.), *The Role of Plant Pathology in Food Safety and Food Security*, Plant Pathology in the 21st Century 3,
DOI 10.1007/978-1-4020-8932-9_9,

F. proliferatum was first described as *Cephalosporium proliferatum* in 1971 but has been recognized as a distinct species within *Fusarium* since 1976 (Nirenberg 1976; Leslie and Summerell 2006). Prior to this, it is likely that the fungus was among the *F. moniliforme* isolates reported from a variety of hosts. However, it is not possible to draw definitive conclusions about the occurrence of the fungus from these reports because most of the isolates are not available for identification by more up-to-date criteria. *F. proliferatum* can be distinguished from related species by morphology, unique DNA sequences, secondary metabolite profile and sexual compatibility. The teleomorph (sexual state) of the fungus was initially described as *Gibberella fujikuroi* variety *intermedium*, subsequently as *G. fujikuroi* mating population D, but more recently as *G. intermedia* (Kuhlman 1982; Leslie 1995; Samuels et al. 2001). *F. proliferatum/G. intermedia* is part of the *Gibberella fujikuroi* species complex, which currently consists of over 40 phylogenetically distinct lineages, including species resolved from *F. moniliforme* and other recently described species (Leslie 1995; Leslie and Summerell 2006; O'Donnell et al. 2000). DNA-based phylogenetic analyses resolved the complex into three distinct clades that do not necessarily correlate with morphology (O'Donnell et al. 2000; Nirenberg and O'Donnell 1998). The analysis also indicated that *F. proliferatum* is most closely related to *F. concentricum*, *F. fujikuroi*, and *F. globosum* but less closely related to (i.e. in a different clade from) *F. nygamai*, *F. subglutinans* and *F. verticillioides*, even though all of these species are morphologically similar.

A critical issue in studies on occurrence, host range and geographical distribution of *F. proliferatum* is whether field isolates of the fungus have been identified correctly. Using morphology to distinguish between *F. proliferatum* and some of the other species within the *G. fujikuroi* species complex is difficult because of their morphological similarities. The morphological characters that distinguish *F. proliferatum* and *F. verticillioides*, for example, are somewhat subtle: *F. verticillioides* produces relatively long chains of microconidia from monophialides, whereas *F. proliferatum* produces shorter chains of microconidia from monophialides and polyphialides, but the polyphialides occur less often (Leslie and Summerell 2006). Identification of *F. proliferatum* by sexual mating assay, developed to induce formation of the teleomorph of heterothallic *Gibberella* species in vitro, can overcome some limitations of morphology-based identification. In the assay, unknown isolates are crossed with well defined, sexually fertile strains, known as mating-type testers. *F. proliferatum* isolates should produce abundant sexual fruiting bodies (perithecia) with ascospores when crossed with tester strains of *F. proliferatum/G. intermedia* but no or rare perithecia when crossed with tester strains of other species (Leslie 1995). The sexual mating assay also has limitations; some isolates of *F. proliferatum* are not sexually fertile and others are interfertile with *F. fujikuroi* (teleomorph *G. fujikuroi*) and as a result produce ascospore-bearing perithecia when crossed with its tester strains (Desjardins et al. 1997; Leslie et al. 2007). *F. proliferatum-F. fujikuroi* hybrid strains have been isolated from native prairie grass in Kansas (Leslie et al. 2004). Thus, hybridization between these species can occur in the field as well as the laboratory.

Perhaps the most reliable method to distinguish between *F. proliferatum* and closely related species is DNA sequence comparisons. DNA used for this has included the nuclear ribosomal DNA intergenic spacer (IGS), the nuclear ribosomal DNA internal transcribed spacer and genes encoding the translation elongation factor 1α (TEF), β-tubulin, calmodulin, cytochrome P450 reductase, and 28S ribosomal RNA (O'Donnell et al. 2000; Tsavkelova et al. 2008; Rim et al. 2005). TEF and IGS are particularly useful because they are the basis of the system used to identify *Fusarium* species via the Fusarium Database (http://isolate.fusariumdb.org/index.php) (Geiser et al. 2004). The database uses the BLASTN algorithm to compare the TEF or IGS sequence of an unknown to TEF or IGS sequences from all species in the database. The results of the analysis indicate to which sequence in the database the unknown sequence is most identical and thereby provide evidence of the species identity of the unknown. A potential limitation of analyzing a single gene to identify an unknown isolate of *F. proliferatum* is the existence of *F. proliferatum* hybrids. However, it should be possible to overcome this limitation by sequencing multiple loci.

At least three species-specific PCR assays have been described for *F. proliferatum* as part of efforts to develop rapid, sensitive, and cost-effective methods for detection and identification of *Fusarium* species. Primers for these assays were designed based on interspecies differences in the calmodulin gene (Mulè et al. 2004) and on anonymous DNA sequences (Dowd et al. 2004; Naef and Défago 2006). Although their initial descriptions reported the assays to be effective, it is not clear how reliable the assays are over the entire range of genetic diversity of *F. proliferatum*. If identification of *F. proliferatum* based on sequence of a single gene can sometimes be insufficient to distinguish it from other species or hybrids, it is likely that PCR assays based on a single primer pair from one gene will not always suffice. PCR assays that employ multiple primer pairs and amplify fragments from multiple loci should overcome limitations of assays based on a single primer pair.

O'Donnell et al. (2007) have recently developed a rapid, multiplex primer extension assay to identify *F. proliferatum* and other *Fusarium* species among isolates from humans. The assay exploited single nucleotide polymorphisms (SNPs) among *Fusarium* species in the RNA polymerase II gene and combined the SNPs with microsphere and laser technology. Although the assay required technology that was expensive compared to more traditional PCR approaches, it will likely be a valuable contribution to the arsenal of tools for identifying and detecting *F. proliferatum* as well as other *Fusarium* species.

9.2 Pathogenicity

F. proliferatum has a remarkably broad host range and has been isolated from at least 25 plant species, including multiple dicot genera, at least two conifer genera and widely grown monocot crops such as maize, rice and wheat (Table 9.1). Although it has been isolated from many plant species, a survey of the literature indicates that its ability to cause disease has been confirmed on about only half of them (Table 9.1). Diseases caused by *F. proliferatum* include blights, diebacks, rots

Table 9.1 Plants from which *Fusarium proliferatum* has been isolated

Plant	Botanical Name	Identification[a]	Pathogenicity Confirmed[b]	Disease[c]	References
Asparagus	*Asparagus officinalis*	Mating	Yes	Crown and root rot	Corpas-Hervias et al. 2006; Elmer 1995; Logrieco et al. 1998
Banana	*Musa* species	Morphology	Yes	Crown rot	Anthony et al. 2004; Jiménez et al. 1997
Chinese Lantern	*Physalis alkekengi*	ITS	n.r.	n.r.	Rim et al. 2005
Citrus	*Citrus reticulata*	Morphology	Yes	Fruit rot	Hyun et al. 2000
Cowpea	*Vigna unguiculata*	Morphology	n.r.	n.r.	Kritzinger et al. 2003
Date Palm	*Phoenix dactylifera*	28S rDNA/ mating	Yes	Wilt and dieback	Abdalla et al. 2000
Douglas Fir	*Pseudotsuga menziesii*	Morphology	Yes	Seedling blight	Dumroese et al. 1998
Fig	*Ficus carica*	Mating	n.r.	Fruit rot	Moretti et al. 2000
Garlic	*Allium sativum*	Mating	Yes	Bulb rot	Dugan et al. 2003; Stankovic et al. 2007
Maize/Corn	*Zea mays*	Mating	Yes	Ear and stalk rot	Desjardins et al. 2000a; Ghiasian et al. 2005; Logrieco et al. 1995; Munkvold et al. 1998
Mango	*Mangifera indica*	Mating	n.r.	Malformation	Leslie 1995
Okra	*Hibiscus esculentum*	Mating	n.r.	n.r.	Kuhlman 1982
Onion	*Allium cepa*	TEF/mating	Yes	Bulb rot	Galván et al. 2008; Stankovic et al. 2007
	Allium species				
Orchid	*Dendrobium moschatum*	ITS IGS CPR TEF TUB	n.r.	No disease	Tsavkelova et al. 2008
Ornamental Palm	*Phoenix canariensis*	Mating	Yes	Wilt and dieback	Armengol et al. 2005

	Chamaerops humilis				
	Washingtonia filifera				
	Washingtonia robusta				
Pearl Millet	*Pennisetum glaucum*	Mating	n.r.	n.r.	Jurjevic et al. 2005
Pine	*Pinus* pinea	28S rDNA	n.r.	n.r.	Marín et al. 2006
	Pinus strobus	Morphology	Yes	Root rot	Ocamb et al. 2002
Prairie Grasses	*Andropogon gerardii*	Mating	n.r.	n.r.	Leslie et al. 2004
	Andropogon scoparius				
	Sorghastrum nuttans				
Giant Reed	*Arundo donax*	Mating	n.r.	n.r.	Láday et al. 2004
Rice	*Oryza sativa*	Mating	n.r.	Bakanae	Desjardins et al. 1997, 2000b
Sorghum	*Sorghum bicolor*	Mating	Yes	Stalk rot	Tesso et al. 2004; Leslie 1995
Sugar Cane	*Saccharum officinarum*	Mating	n.r.	n.r.	Kuhlman 1982
Tomato	*Lycopersicon esculentum*	Mating	n.r.	Wilt	Moretti et al. 1999
Wheat	*Triticum aestivum*	TEF/mating	Yes	Black point	Conner et al. 1996; Desjardins et al. 2007

[a]Morphology indicates that the identity of *F. proliferatum* was determined by examination of colony morphology only. Mating indicates that the morphological identification was confirmed by mating analysis with standard *G. intermedia* strains. 28S rDNA, ITS, IGS, CPR, TEF and TUB indicate that the identity was confirmed by DNA sequence analysis. The sequenced DNA was: 28S ribosomal RNA gene (28S rDNA), cytochrome P450 reductase gene (CPR), nuclear ribosomal DNA intergenic spacer (IGS), nuclear ribosomal DNA internal transcribed spacer (ITS), translation elongation factor 1α gene (TEF), β-tubulin gene (TUB).

[b]n.r. = not reported in references listed.

[c]Disease caused by *F. proliferatum* on the plant indicated.

and wilt. Rots can occur on roots, bulbs, crowns, stems, shoots, fruits and seeds. *F. proliferatum* can also grow without causing symptoms (i.e. endophytically) in at least some hosts, including maize, orchids and wheat (Jeney et al. 2007; Kwon et al. 2001; Tsavkelova et al. 2008).

F. proliferatum and morphologically similar *Fusarium* species often cause similar disease symptoms on the same plant species. *F. proliferatum* and *F. oxysporum* are considered the major causal agents of asparagus crown and root rot (Corpas-Hervias et al. 2006; Elmer et al. 1996). Although *F. fujikuroi* (teleomorph *G. fujikuroi*) is regarded as the primary cause of bakanae disease of rice, *F. proliferatum* also occurs on rice and may contribute to the disease (Desjardins et al. 1997, 2000b). The symptomology of bakanae disease is complex, but perhaps the most recognized symptom is seedling elongation that results from gibberellin production by *Fusarium*. Because gibberellin production is rare in *F. proliferatum* but common in *F. fujikuroi* (Malonek et al. 2005a), *F. proliferatum* is unlikely to contribute significantly to seedling elongation. Although *F. verticillioides* is considered the major cause of Fusarium ear rot (pink ear rot) of maize, *F. proliferatum* and *F. subglutinans* can also cause this disease. The prevalence of *F. proliferatum* in maize appears to vary with environmental conditions and or geographic regions. In surveys of maize from the Iowa (USA), Italy, Nepal, and Slovakia, *F. proliferatum* was recovered from 100%, 83%, 4%, and 54%, respectively, of samples (Desjardins et al. 2000a; Logrieco et al. 1995; Munkvold et al. 1998; Srobarova et al. 2002). When the incidence of species has been compared, *F. proliferatum* is generally recovered at lower frequencies than *F. verticillioides*. In the survey of Nepalese maize, of 103 isolates that had a morphology consistent with the *G. fujikuroi* species complex, 80% were *F. verticillioides* and 5% were *F. proliferatum* based on mating tests (Desjardins et al. 2000a). In the survey of visibly diseased maize from Italy, morphological identification indicated that *F. verticillioides* was isolated from an average of 54% of kernels and *F. proliferatum* from 34% (Logrieco et al. 1995).

There are also reports that *F. proliferatum* is pathogenic on organisms other than plants. *F. proliferatum* has been observed growing inside and around sporangiophores of *Plasmopara viticola*, the oomycete that causes grape downy mildew, and application of *F. proliferatum* can reduce the severity of downy mildew on grape leaves (Falk et al. 1996). *F. proliferatum* has insecticidal activity against and can colonize bodies of the aphid *Schizaphis graminum* (Ganassi et al. 2000). *F. proliferatum* has also been isolated from humans with fungal infections (O'Donnell et al. 2007).

The broad host range and geographic distribution of *F. proliferatum* raise the question of whether the species has genetic substructure based on host or geographic origin. Namely, are strains that occur in one host or in one geographic region genetically more similar to each other than to strains from other hosts or regions? To address this question, we conducted a DNA-based phylogenetic analysis of *F. proliferatum*. We selected 50 isolates of the fungus recovered from seven different plants and three continents (Fig. 9.1). For comparison we also selected several isolates of *F. fujikuroi* and *F. verticillioides*. We amplified and determined the nucleotide sequence of fragments of four nuclear genes, aligned the sequence data, and the subjected their to maximum parsimony analysis. The results of the analysis indicate considerable genetic variation among the 50 isolates of *F. proliferatum*

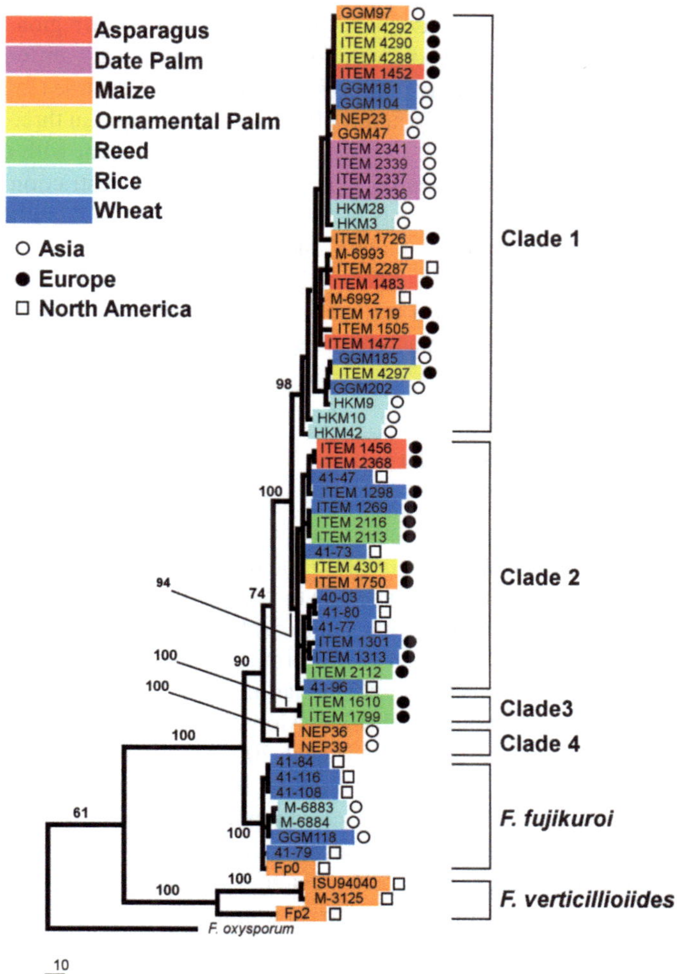

Fig. 9.1 Phylogenetic tree resulting from maximum parsimony analysis of *F. proliferatum* isolates from seven plant species. The 50 *F. proliferatum* strains analyzed are included in clades 1–4. Sequence data for multiple strains of *F. fujikuroi* and *F. verticillioides* were included for comparison. Data from a single strain of *F. oxysporum* f. sp. *lycopercisi* was used as the outgroup. The analysis employed combined DNA sequence data from PCR-amplified fragments of four genes: (1) cytochrome p450 NADPH-dependent reductase gene *CPR1/cpr1*; (2) histone H3 gene; (3) RNA polymerase II gene *RPB2*; and (4) translation elongation factor 1-α gene *TEF1*. The total number of nucleotides analyzed was 3,194. The number of parsimony-informative characters was 260. The analysis generated >100 trees. Bootstrap values are based on 1,000 pseudoreplications. The sequence alignments were done with the program ClustalW and the parsimony and bootstrap analyses were done with PAUP version 4.0b10 for Unix

but resolved all of them into four strongly supported clades (bootstrap values ≥94) that were distinct from *F. fujikuroi* and *F. verticillioides* clades (Fig. 9.1). The results did not provide evidence for substructure based on host or geographic origin. Isolates of *F. proliferatum* from different hosts and geographic regions were sometimes as genetically similar as isolates from the same host or region. Likewise, isolates from the same host or region were sometimes as genetically different as isolates from different hosts or regions. This was exemplified by Nepalese maize isolates (GGM47, GGM97, NEP23, NEP36 and NEP39), which occurred in three out of the four *F. proliferatum* clades (Fig. 9.1). These results are consistent with other phylogenetic analyses of *F. proliferatum* that employed either restriction fragment length polymorphism (RFLP) analysis of mitochondrial DNA (Láday et al. 2004) or nucleotide sequence analysis of three nuclear loci (Geiser et al. 2003). Our analysis and that of (Láday et al. 2004) did not include isolates recovered from animals or dicotyledonous plants. Thus, we cannot rule out the possibility that there is genetic substructure among isolates associated with these hosts (Fig. 9.2).

9.3 Secondary Metabolism

F. proliferatum can produce biologically active secondary metabolites that are diverse in structure and biosynthetic origins. The metabolites include the polyketide pigment bikaverin, the terpenoid plant growth regulators gibberellic acids (GAs), and multiple mycotoxins, including the cyclic peptide beauvericin, the polyketides fumonisins, fusaric acid, fusarins, and moniliformin, and the terpenoid fusaproliferin (Armengol et al. 2005; Bacon et al. 1996; Desjardins et al. 1997, 2000b; Leslie et al. 2004; Marasas et al. 1986; Miller et al. 1995; Moretti et al. 1994, 1997; Plattner and Nelson 1994; Ritieni et al. 1995; Stankovic et al. 2007). Fumonisins receive the most attention because they are common contaminants in maize grain and there are epidemiological associations between the consumption of fumonisin-contaminated maize and esophageal cancer as well as neural tube defects in some human populations (Marasas 2001; Suarez et al. 2000). Fumonisins can also cause cancer and neural tube defects in laboratory rodents, brain lesions (leukoencephalomalacia) in horses and lung edema in swine (Marasas et al. 2004). Although *F. proliferatum* can cause ear rot of maize, its contribution to fumonisin contamination in this crop is not always clear because of the higher incidence of the related fumonisin-producing fungus, *F. verticillioides*, in maize. Nevertheless, *F. proliferatum* can be similar to *F. verticillioides* in the levels and kinds of fumonisins it produces in culture

Fig. 9.2 Chemical structures of biologically active secondary metabolites produced by *F. proliferatum*. Gibberellin A4 (GA_4) is only one of at least seven gibberellins produced by the fungus. The other gibberellins (GA_1, GA_3, GA_7, GA_9, GA_{20}, and GA_{24}) can differ in structure from GA_4 at multiple positions. *F. proliferatum* can produce fusarins A and F as well as fusarin C, and their structures differ from that of fusarin C in the following ways: fusarin A, the oxygen atom of the terminal hydroxyl is covalently bound to carbon atom 14 to form a five-member ring with the pyrrolidone moiety and the 13,14-epoxide is absent; fusarin F, the epoxide is at carbon atoms 14 and 15 and there is a hydroxyl at carbon atom 13. The structures of the four B-series fumonisins (FB_1, FB_2, FB_3, and FB_4) produced by *F. proliferatum* differ at R_1 and R_2. In FB_1, R_1 and R_2 are OH; in FB_2, R_1 is OH and R_2 is H; in FB_3, R_1 is H and R_2 is OH; and in FB_4, R_1 and R_2 are H

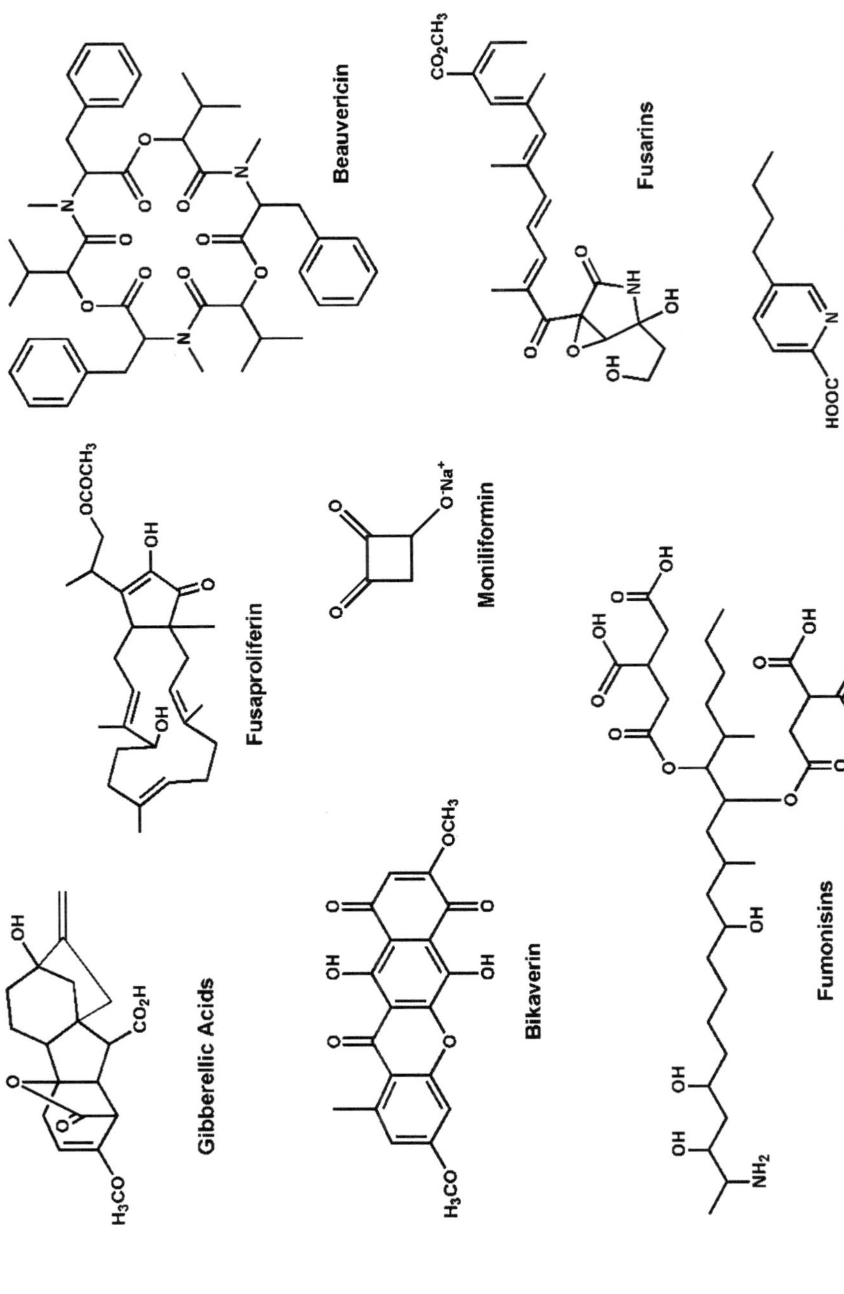
Beauvericin
Fusarins
Fusaric Acid
Fusaproliferin
Moniliformin
Gibberellic Acids
Bikaverin
Fumonisins

(Leslie et al. 2004; Marín et al. 1999; Rheeder et al. 2002; Thiel et al. 1991) and in maize ears (Desjardins and Plattner 2000; Desjardins et al. 2002; Pascale et al. 2002). Beauvericin and fusaproliferin can also contaminate maize inoculated with *F. proliferatum* (Pascale et al. 2002).

Fumonisin contamination that results or likely results from *F. proliferatum* infection has been reported in other crops. In rice infected with *F. proliferatum*, fumonisin B_1 (FB_1) and B_2 (FB_2) were detected at levels of ≤4.3 and ≤1.2 μg/g, respectively, in 40% of samples analyzed (Abbas et al. 1998). The fumonisins were present primarily in rice hulls and to a lesser extent in bran; fumonisins were not detected in rice grain after it was polished. FB_1 was detected in asparagus spears naturally infected with *F. proliferatum* at levels of 0.04–4.5 μg/g in one study (Seefelder et al. 2002) and 0.46–7.4 μg/g in another (Logrieco et al. 1998). FB_1 has also been detected in pine nuts (≤0.05–0.69 μg/g) and garlic bulbs (0.03–0.09 μg/g) inoculated with *F. proliferatum*. In grain from wheat ears inoculated with *F. proliferatum*, the levels of FB_1, FB_2 and FB_3 combined were as high as 140 μg/g for one cultivar–strain combination (Desjardins et al. 2007), but 0.13–65 μg/g for all other cultivar–strain combinations.

A fumonisin biosynthetic gene (*FUM*) cluster has been described in *F. proliferatum* (Waalwijk et al. 2004). The order and orientation of genes within the cluster are the same as in the *FUM* clusters described in *F. verticillioides* and *F. oxysporum* strain O-1890 (Proctor et al. 2003, 2008). The regions flanking the cluster differ in the three species, indicating that the genomic location of the cluster differs in each species. Waalwijk et al. (2004) concluded that the low level of identity (77–89% at amino acid level) of *FUM* genes in *F. proliferatum* and *F. verticillioides* and the different genomic locations of the cluster indicate that each species may have acquired the cluster independently. However, the different genomic locations of the clusters and the discontinuity of *FUM* genes among closely related species (Proctor et al. 2004) could also indicate the cluster was present in ancestral population of *Fusarium* and subsequently moved or was lost during speciation (Proctor et al. 2008).

Although *F. proliferatum* isolates do not generally produce GAs (Desjardins et al. 2000b; Malonek et al. 2005a), at least five GA-producing isolates of *F. proliferatum* have been reported. One was recovered from the orchid *Dendrobium moschatum* (Tsavkelova et al. 2008) and four from Chinese lantern, *Physalis alkekengi* var. *fracheti* (Rim et al. 2005). All GA-producing and nonproducing strains that have been examined in detail have the GA biosynthetic gene cluster. The order and orientation of genes is the same as in the cluster characterized in *F. fujikuroi* (Malonek et al. 2005a). The lack of GA production in *F. proliferatum* can result from reduced expression and mutations in the coding regions of genes in the GA biosynthetic gene cluster (Malonek et al. 2005b, c). Examination of additional isolates of *F. proliferatum* may reveal whether GA production occurs in strains of the fungus recovered from hosts other than D. moschatum or P. alkekengi. Such investigations may also provide insight into the evolutionary forces that have caused loss of GA production in most strains of the fungus but its retention in others.

9.4 Future Research

Understanding the biological and chemical diversity of *F. proliferatum* has been hampered in the past by difficulties in differentiating the fungus from morphologically similar species. During the past three decades, mating type and DNA-base analyses have more clearly delineated *F. proliferatum* from other species. This has provided a greater appreciation for the extraordinarily broad host range and metabolite production of the fungus. This improved understanding of *F. proliferatum* has raised some interesting questions about its biology. For example, what traits allow *F. proliferatum* to be a pathogen of so many and such diverse hosts? In addition, what traits allow for endophytic and or pathogenic growth in at least some hosts, and are these traits the same as those that allow endophytic and pathogenic growth of *F. verticillioides* in maize? The ability to conduct meiotic and molecular genetic (e.g. gene deletion) analyses with *F. proliferatum* make it an excellent system to address such questions. In addition to these fundamental biological questions, there is a need to assess the health risks posed to humans and animals by food and feed derived from crops contaminated with *F. proliferatum* mycotoxins.

Acknowledgement We are grateful to Marcie L. Moore and Stephanie N. Folmar for excellent technical assistance and to Hira K. Manandhar and Gyanu Manandhar for collection of grain samples in Nepal. Partial funding was provided to RHP by the International Short-term Mobility Program, National Research Council, Italy.

References

Abbas H, Cartwright RD, Shier WT, Abouzied MM, Bird CB, Rice LG, Ross PF et al (1998) Natural occurrence of fumonisins in rice with Fusarium sheath rot disease. Plant Dis 82:22–25

Abdalla MY, Al-Rokibah A, Moretti A, Mulè G (2000) Pathogenicity of toxigenic *Fusarium proliferatum* from date palm in Saudi Arabia. Plant Dis 84:321–324

Anthony S, Abeywickrama K, Dayananda R, Wijeratnam SW, Arambewela L (2004) Fungal pathogens associated with banana fruit in Sri Lanka, and their treatment with essential oils. Mycopathologia 157:91–97

Armengol J, Moretti A, Perrone G, Vicent A, Bengoechea JA, García-Jiménez J (2005) Identification, incidence and characterization of *Fusarium proliferatum* on ornamental palms from Spain. Eur J Plant Pathol 112:123–131

Bacon CW, Porter JK, Norred WP, Leslie JF (1996) Production of fusaric acid by *Fusarium* species. Appl Environ Microbiol 62:4039–4043

Conner RL, Hwang SF, Stevens RR (1996) *Fusarium proliferatum*: a new causal agent of black point in wheat. Can J Plant Pathol 18:419–423

Corpas-Hervias C, Melero-Vara JM, Molinero-Ruiz ML, Zurera-Muñoz C, Basallote-Ureba MJ (2006) Characterization of isolates of *Fusarium* spp. obtained from asparagus in Spain. Plant Dis 90:1441–1451

Desjardins AE (2006) Fusarium mycotoxins chemistry, genetics and biology. APS Press, St. Paul, MN

Desjardins AE, Busman M, Proctor RH, Stessman R (2007) Wheat kernel black point and fumonisin contamination by *Fusarium proliferatum*. Food Addit Contam 24:1131–1137

Desjardins AE, Manandhar G, Plattner RD, Maragos CM, Shrestha K, McCormick SP (2000a) Occurrence of *Fusarium* species and mycotoxins in Nepalese maize and wheat and the effect of traditional processing methods on mycotoxin levels. J Agric Food Chem 48:1377–1383

Desjardins AE, Manandhar HK, Plattner RD, Manandhar GG, Poling SM, Maragos CM (2000b) *Fusarium* species from Nepalese rice and production of mycotoxins and gibberellic acid by selected species. Appl Environ Microbiol 66:1020–1025

Desjardins AE, Munkvold GP, Plattner RD, Proctor RH (2002) *FUM1*–a gene required for fumonisin biosynthesis but not for maize ear rot and ear infection by *Gibberella moniliformis* in field tests. Mol Plant-Microbe Interact 15:1157–1164

Desjardins AE, Plattner RD (2000) Fumonisin B_1-nonproducing strains of *Fusarium verticillioides* cause maize (*Zea mays*) ear infection and ear rot. J Agric Food Chem 48:5773–5780

Desjardins AE, Plattner RD, Nelson DR (1997) Production of fumonisin B_1 and moniliformin by *Gibberella fujikuroi* from rice and from various geographic areas. Appl Environ Microbiol 63:1838–1842

Dowd PF, Barnett CJ, Johnson ET, Beck JJ (2004) Leaf axil sampling of midwest U.S. maize for mycotoxigenic *Fusarium* fungi using PCR analysis. Mycopathologia 158:431–440

Dugan FM, Hellier BC, Lupien SL (2003) First report of *Fusarium proliferatum* causing rot of garlic bulbs in North America. Plant Pathol 52:426

Dumroese RK, James RL, Wenny DL (1998) Interactions among *Streptomyces griseoviridis*, *Fusarium* root disease, and Douglas-fir seedlings. New Forest 15:181–191

Elmer WH (1995) A single mating population of *Gibberella fujikuroi* (*Fusarium proliferatum*) predominates in asparagus fields in Connecticut, Massachusetts, and Michigan. Mycologia 87:68–71

Elmer WH, Johnson DA, Mink GI (1996) Epidemiology and management of the diseases causal to asparagus decline. Plant Dis 80:117–125

Falk SP, Pearson RC, Gadoury DM, Seem RC, Sztejnberg A (1996) *Fusarium proliferatum* as a biocontrol agent against the grape downy mildew. Phytopathology 86:1010–1017

Galván GA, Koning-Boucoiran FS, Koopman WJM, Burger-Meijer K, Gozález PH, Waalwijk C, Kik C et al (2008) Genetic variation among *Fusarium* isolates from onion, and resistance to *Fusarium* basal rot in related *Allium* species. Eur J Plant Pathol 121:499–512

Ganassi S, Moretti A, Stornelli C, Fratello B, Bonvincini Pagliai AM, Logrieco A, Sabatini MA (2000) Effect of *Fusarium*, *Paecilomyces* and *Trichoderma* formulations against aphid *Schizaphis graminum*. Mycopathologia 151:131–138

Geiser DM, Abbas HK, Desjardins AE, Hackett M, Juba JH, O'Donnell KL, Royse J, Tunali B (2003) Phylogenetic and biological species boundaries around *Fusarium proliferatum*. Proceedings of the Mycological society of America, Pacific Grove, CA

Geiser DM, del Mar Jiménez-Gasco M, Kang S, Makalowska I, Veeraraghavan N, Ward TJZN, Kuldau GA et al (2004) FUSARIUM-ID v. 1.0: A DNA sequence database for identifying *Fusarium*. Eur J Plant Pathol 110:473–479

Ghiasian SA, Rezayat SM, Kord-Bacheh P, Maghsood AH, Yazdanpanah H, Shepard GS, Van Der Westhuizen L et al (2005) Fumonisin production by *Fusarium* species isolated from freshly harvested corn in Iran. Mycopathologia 159:31–40

Hyun JW, Lee S-C, Kim D-H, Ko S-W, Kim K-S (2000) *Fusarium* fruit rot of citrus in Jeju Island. Mycobiology 28:158–162

Jeney A, Béki E, Keszthelyi A, Leslie JF, Hornok L (2007) Cloning and characterization of *Fpmtr1*, an amino acid transporter gene of *Fusarium proliferatum* (*Gibberella intermedia*). J Basic Microbiol 47:16–24

Jiménez M, Huerta T, Mateo R (1997) Mycotoxin production by *Fusarium* species isolated from bananas. Appl Environ Microbiol 63:364–369

Jurjevic Z, Wilson DM, Wilson JP, Geiser DM, Juba JH, Mubatanhema W, Widstrom NW et al (2005) *Fusarium* species of the *Gibberella fujikuroi* complex and fumonisin contamination of pearl millet and corn in Georgia, USA. Mycopathologia 159:401–406

Klittich CJR, Leslie JF, Nelson PE, Marasas WFO (1997) *Fusarium thapsinum* (*Gibberella thapsina*): a new species in section *Liseola* from sorghum. Mycologia 89:643–652

Kritzinger Q, Aveling TAS, Marasas WFO, Rheeder JP, Van Der Westhuizen L, Shepard GS (2003) Mycoflora and fumonisin mycotoxins associated with cowpea [*Vigna unguiculata* (L.) Walp] seeds. J Agric Food Chem 51:2188–2192

Kuhlman EG (1982) Varieties of *Gibberella fujikuroi* with anamorphs in *Fusarium* section *Liseola*. Mycologia 74:759–768

Kwon S-I, von Dohlen CD, Anderson AJ (2001) Gene sequence analysis of an opportunistic wheat pathogen, an isolate of *Fusarium proliferatum*. Can J Bot 79:1115–1121

Láday M, Mulè G, Moretti A, Juhász Á, Szécsi Á, Logrieco A (2004) Mitochondrial DNA variability in *Fusarium proliferatum* (*Gibberella intermedia*). Eur J Plant Pathol 110:563–571

Leslie JF (1995) *Gibberella fujikuroi*: available populations and variable traits. Can J Bot 73:S282–S291

Leslie JF, Anderson LL, Bowden RL, Lee Y-W (2007) Inter- and intra-specific genetic variation in *Fusarium*. Int J Food Microbiol 119:25–32

Leslie JF, Summerell BA (2006) The *Fusarium* Laboratory Manual. Blackwell Publishing, Ames, IA

Leslie JF, Zeller KA, Logrieco A, Mulè G, Moretti A, Ritieni A (2004) Species diversity of and toxin production by *Gibberella fujikuroi* species complex strains isolated from native prairie grasses in Kansas. Appl Environ Microbiol 70:2254–2262

Logrieco A, Doko B, Moretti A, Frisullo S, Visconti A (1998) Occurrence of fumonisin B_1 and B_2 in *Fusarium proliferatum* infected asparagus plants. J Agric Food Chem 46:5201–6204

Logrieco A, Moretti A, Retieni A, Bottalico A, Corda P (1995) Occurrence and toxigenicity of *Fusarium proliferatum* from preharvest maize ear rot, and associated mycotoxins, in Italy. Plant Dis 79:727–731

Malonek S, Bömke C, Bornberg-Bauer E, Rojas MC, Hedden P, Hopkins P, Tudzynski B (2005a) Distribution of gibberellin biosynthetic genes and gibberellin production in the *Gibberella fujikuroi* species complex. Phytochemistry 66:1296–1311

Malonek S, Rojas MC, Hedden P, Gaskin P, Hopkins P, Tudzynski B (2005b) Functional characterization of two cytochrome P450 monooxygenase genes, *P450-1* and *P450-4*, of the gibberellic acid gene cluster in *Fusarium proliferatum* (*Gibberella fujikuroi* MP-D). Appl Environ Microbiol 71:1462–1472

Malonek S, Rojas MC, Hedden P, Hopkins P, Tudzynski B (2005c) Restoration of gibberellin production in *Fusarium proliferatum* by functional complementation of enzymatic blocks. Appl Environ Microbiol 71:6014–6025

Marasas WFO (2001) Discovery and occurrence of the fumonisins: a historical perspective. Environ Health Perspect 109(2):239–243

Marasas WFO, Nelson PE, Toussoun TA (1984) Toxigenic *Fusarium* species: identity and mycotoxicology. The Pennsylvania State University Press, University Park, PA

Marasas WFO, Riley RT, Hendricks K, Stevens VL, Sadler TW, Gelineau-van Waes J, Missmer SA et al (2004) Fumonisins disrupt shingolipid metabolism, folate transport and neural tube development in embryo culture and in vivo: a potential risk factor for human neural tube defects among populations consuming fumonisin-contaminated maize. J Nutri 134:711–716

Marasas WFO, Thiel PG, Rabie CJ (1986) Moniliformin production in *Fusarium* section *Liseola*. Mycologia 78:242–247

Marín S, Magan N, Serra J, Ramos AJ, Canela R, Sanchis V (1999) Fumoinisin B1 production and growth of *Fusarium moniliforme* and *Fusarium proliferatum* on maize, wheat, and barley grain. J Food Sci 64:921–924

Marín S, Ramos AJ, Vásquez C, Sanchis V (2006) Contamination of pine nuts by fumonisin produced by strains of *Fusarium prolieratum* isolated from *Pinus pinea*. Lett Appl Microbiol 44:68–72

Miller JD, Savard ME, Schaafsma AW, Seifert KA, Reid LM (1995) Mycotoxin production by *Fusarium moniliforme* and *Fusarium proliferatum* from Ontario and occurrence of fumonisins in the 1993 corn crop. Can J Plant Pathol 17:233–239

Moretti A, Ferracane R, Ritieni A, Frisullo S, Lops A, Logrieco A (2000) *Fusarium* species from fig in Apulia: biological and toxicological characterization. Mitt Biol Bundesanst Land-Forstwirtsch Berlin-Dahlem 377:31–32

Moretti A, Logrieco A, Bottalico A, Ritieni A, Fogliano V, Randazzo G (1997) Diversity in beauvericin and fusaproliferin production by different populations of *Gibberella fujikuroi* (*Fusarium* section *Liseola*). Sydowia 48:44–56

Moretti A, Logrieco A, Bottalico A, Ritieni A, Randazzo G (1994) Production of beauvericin by *Fusarium proliferatum* from maize in Italy. Mycotoxin Res 10:73–78

Moretti A, Mulè G, Perrone G, Bottalico A, D'erchia AM, Logrieco A (1999) *Fusarium proliferatum* from various plants: fertility, toxigenicity and characterization by RAPDs. Bull Inst Compr Agr Sci Kinki Univ 7:27–36

Mulè G, Susca A, Stea G, Moretti A (2004) A species-specific PCR assay based on the calmodulin partial gene for identification of *Fusarium verticillioides, F. proliferatum* and *F. subglutinans*. Eur J Plant Pathol 110:495–502

Munkvold G, Stahr HM, Logrieco A, Moretti A, Ritieni A (1998) Occurrence of fusaproliferin and beauvericin in *Fusarium*-contaminated livestock feed in Iowa. Appl Environ Microbiol 64:3923–3926

Naef A, Défago G (2006) Population structure of plant-pathogenic *Fusarium* species in overwintered stalk residues from Bt-transformed and non-transformed maize crops. Eur J Plant Pathol 116:129–143

Nelson PE, Toussoun TA, Marasas WFO (1983) *Fusarium* species: An Illustrated Manual for Identification. The Pennsylvania State University Press, University Park, PA

Nirenberg HI (1976) Untersuchugen über die morphologische and biologische Differenzierung in der *Fusarium*-Sektion *Liseola*. Mitt Biol Bundesanst Land-Forstwirtsch Berlin-Dahlem 169:1–117

Nirenberg HI, O'Donnell K (1998) New *Fusarium* species and combinations within the *Gibberella fujikuroi* species complex. Mycologia 90:434–458

O'Donnell K, Nirenberg HI, Aoki T, Cigelnik E (2000) A multigene phylogeny of the *Gibberella fujikuroi* species complex: detection of additional phylogenetically distinct species. Mycoscience 41:61–78

O'Donnell K, Sarver BAJ, Brandt M, Chang DC, Noble-Wang J, Park BJ, Sutton DA et al (2007) Phylogenetic diversity and microsphere array-based genotyping of human pathogenic Fusaria, including from the multistate contact lens-associated U.S. keratitis outbreaks of 2005 and 2006. J Clin Microbiol 45:2235–2248

Ocamb CM, Juzwik J, Martin FB (2002) *Fusarium spp.* and *Pinus strobus* seedlings: root disease pathogens and taxa associated with seed. New Forest 24:67–79

Pascale M, Visconti A, Pronczuk M, Wisniewska H, Chelkowski J (2002) Accumulation of fumonisins, beauvericin and fusaproliferin in maize hybrids inoculated under field conditions with *Fusarium proliferatum*. Mycol Res 106:1026–1030

Plattner RD, Nelson PE (1994) Production of beauvericin by a strain of *Fusarium proiferatum* isolated from corn fodder for swine. Appl Environ Microbiol 60:3894–3896

Proctor RH, Brown DW, Plattner RD, Desjardins AE (2003) Co-expression of 15 contiguous genes delineates a fumonisin biosynthetic gene cluster in *Gibberella moniliformis*. Fungal Genet Biol 38:237–249

Proctor RH, Busman M, Seo J-A, Lee Y-W, Plattner RD (2008) A fumonisin biosynthetic gene cluster in *Fusarium oxysporum* strain O-1890 and the genetic basis for B versus C fumonisin production. Fungal Genet Biol 45:1016–1026

Proctor RH, Plattner RD, Brown DW, Seo J-A, Lee Y-W (2004) Discontinuous distribution of fumonisin biosynthetic genes in the *Gibberella fujikuroi* species complex. Mycol Res 108:815–822

Rheeder JP, Marasas WFO, Vismer HF (2002) Production of fumonisin analogs by *Fusarium* species. Appl Environ Microbiol 68:2101–2105

Rim S-O, Lee J-H, Choi W-Y, Hwang S-K, Suh S-J, Lee I-J, Rhee I-K et al (2005) *Fusarium proliferatum* KGL0401 as a new gibberellin-producing fungus. J Microbiol Biotechnol 15:809–814

Ritieni A, Fogliano V, Randazzo G, Scarallo A, Logrieco A, Moretti A, Mannina L et al (1995) Isolation and characterization of fusaproliferin, a new toxic metabolite from *Fusarium proliferatum*. Nat Toxins 3:17–20

Samuels GJ, Nirenberg HI, Seifert KA (2001). In: Summerell B, Leslie JF, Backhouse D, Bryden WL, Burgess LW (eds) Perithecial species of *Fusarium*. *Fusarium*: Paul E. Nelson Memorial Symposium. APS Press, St. Paul, MN, pp 1–14

Seefelder A, Gossman M, Humpf H-U (2002) Analysis of fumonisin B1 in *Fusarium proliferatum*-infected asparagus spears and garlic bulbs from Germany by liquid chromatography-electrospray ionization mass spectrometry. J Agric Food Chem 50:2778–2781

Sheldon JL (1904) A corn mold (*Fusarium moniliforme* n. sp.). Agric Exp Sta Nebr Annu Rep 17:32

Srobarova A, Moretti A, Ferracane R, Ritieni A, Logrieco A (2002) Toxigenic *Fusarium* species of *Liseola* section in pre-harvest maize ear rot, and associated mycotoxins in Slovakia. Eur J Plant Pathol 108:299–306

Stankovic S, Levic J, Petrovic T, Logrieco A, Moretti A (2007) Pathogenicity and mycotoxin production by *Fusarium proliferatum* isolated from onion and garlic in Serbia. Eur J Plant Pathol 118:165–172

Suarez L, Hendricks KA, Cooper SP, Sweeny AM, Hardy RJ, Larsen RD (2000) Neural tube defects among Mexican Americans living on the US-Mexico border: effects of folic acid and dietary folate. Amer J Epidemiol 152:1017–1023

Tesso T, Claflin LE, Tuinstra MR (2004) Estimation of combining ability for resistance to Fusarium stalk rot in grain Sorghum. Crop Sci 44:1195–1199

Thiel PG, Marasas WFO, Sydenham EW, Shepard GS, Gelderblom WCA, Nieuwenhuis JJ (1991) Survey of fumonisin production by *Fusarium* species. Appl Environ Microbiol 57:1089–1093

Tsavkelova EA, Bömke C, Netrusov AI, Weiner J, Tudzynski B (2008) Production of gibberellic acids by an orchid-associated *Fusarium proliferatum* strain. Fungal Genet Biol 45:1393–1403

Waalwijk C, van der Lee T, de Vries I, Hesselink T, Arts J, Kema GHJ (2004) Synteny in toxigenic *Fusarium* species: the fumonisin gene cluster and the mating type region as examples. Eur J Plant Pathol 110:533–544

Part IV
Biosecurity and Quarantine

Chapter 10
Bioterrorism: A Threat to Plant Biosecurity?

J.P. Stack, F. Suffert, and M.L. Gullino

10.1 Plant Biosecurity: The Foundation for Food Security

10.1.1 What Is Plant Biosecurity?

There are many definitions of biosecurity ranging from very broad to very narrow in nature and scope. Much of the variation can be attributed to the scale of the systems under consideration and the attributes being analyzed. For the purpose of this paper, biosecurity will be used as previously defined (Stack 2008). Security is a state of existence that assures safety and provides protection from harm; real or perceived. Biosecurity is in specific reference to protection from harm caused by biological agents. Biosecurity at the laboratory scale is focused on physical and behavioral measures that ensure specific organisms cannot accidentally escape or be deliberately taken from the laboratory. Geographic biosecurity is focused on exclusion and containment; it is about ensuring that exotic organisms are not introduced into a given area and that potentially harmful organisms cannot escape from that area. Plant biosecurity is concerned with the protection of natural and managed plant systems from the introduction of exotic organisms or from the emergence of indigenous organisms that would negatively impact the productivity, sustainability,

J.P. Stack
Department of Plant Pathology, Kansas State University,
4024 Throckmorton Plant Sciences Center, Manhattan, Kansas, 66506, USA
e-mail: jstack@ksu.edu

F. Suffert
AgroParisTech, Institut National de la Recherche Agronomique (INRA), UMR 1290
BIOGER-CPP, F-78850, Thiverval-Grignon, France
e-mail: fsuffert@grignon.inra.fr

M.L. Gullino
Centre of Competence for the Innovation in the Agro-environmental Sector, University of Torino,
Via Leonardo da Vinci, 44, 10095, Grugliasco (Torino), Italy
e-mail: marialodovica.gullino@unito.it

R.N. Strange and M.L. Gullino (eds.), *The Role of Plant Pathology in Food Safety and Food Security,* Plant Pathology in the 21st Century 3,
DOI 10.1007/978-1-4020-8932-9_10, © Springer Science+Business Media B.V. 2010

or diversity of plant systems. Plant biosecurity is a state of preparedness that ensures productive and sustainable plant ecosystems. In agricultural systems, plant biosecurity is a state of preparedness that ensures a safe and constant supply of food, feed, fiber, timber and fuels. Crop biosecurity has also been defined as protecting a state from invasive plant pathogens (Brasier 2008).

10.1.2 The Food Security – Plant Biosecurity Linkage

The food systems that provide the caloric requirements for most of the world's population are plant-based, including, rice, wheat and maize. We either consume those plants directly or we provide them as feed, forage, or grazing to the animals that we consume. The health and productivity of plant systems are prerequisites for food security and human health. At present, many nations lack the food production capacity to feed their existing and projected populations. They are dependent upon international aid programs and trade of plants and plant products to compensate for food deficits. Consequently, global food security is, in part, dependent upon effective plant biosecurity strategies and appropriate infrastructure at the national and international levels (Stack and Fletcher 2007). In food production systems, plant biosecurity is a state of preparedness that ensures a safe, affordable, and available supply of food and feed. Food protests and riots in at least 30 nations over the last 18 months are evidence of the significant linkage between food security and national security (Shelburne 2008). Without effective plant biosecurity programs to protect the world's staple crops, food safety and security will decline in developing nations and weak governments will fail (Shelburne 2008). This will further compromise global economic development and international programs to reduce hunger and improve health. Without effective plant biosecurity programs to protect the world's natural plant systems, the ecosystem services that they provide to support humans will decline, thus compromising the development of sustainable societies.

10.2 Threats to Plant Biosecurity

10.2.1 General Threats to Plant Biosecurity

There are many general threats to plant systems that put plant biosecurity at risk, including global trade of plants and plant products, climate change, population growth and landscape exploitation (Stack 2008; Gullino et al. 2008; Brasier 2008). One of the significant difficulties associated with addressing the threats of bioterrorism and biocrime targeting plant systems is the inability to determine that an outbreak was intentionally caused (Fletcher 2008). There is a lack of specific criteria by which to distinguish intentional from accidental or natural introductions (Nutter and Madden 2005; Fletcher et al. 2006). This is made much more difficult by the increasing frequency of accidental introductions associated with global plant trade (Palm 1999; Rossman 2001; Brasier 2008). Natural dispersal of plants via ocean currents or air systems is insignificant compared to the intentional movement of plants for the purpose of global trade (Mack and Lonsdale 2001). The deliberate trade of ornamental and landscape plants began and continues without adequate regard for the consequences of those introductions to the new environments (Britton 2004; Brasier 2008). Plant pathogens can be readily dispersed within plants and plant products in a latent, asymptomatic phase. The number of plant pathogens introduced into the United States increased approximately 300% from 1900 to 2000 (Windle 2004). Whether the increasing numbers are due solely to new introductions or a combination of new introductions and better detection protocols and diagnostic technologies is difficult to determine. However, a recent United States Department of Agriculture (USDA) estimate indicates that a new exotic species is detected in the U.S. every 8–12 days. This large-scale influx of exotic species has been termed biological pollution (Britton 2004). With that as the norm, determining a specific introduction to be intentional will be challenging. Further complicating efforts to assign attribution, global trade in plants has facilitated the evolution of new pathogen species with novel properties that would not be predicted from the parental phenotypes (Man in't Veldt et al., 1998; Brasier 2001). Newly described species have been detected in nurseries where plants from geographically distinct areas of the world were placed in close proximity to each other. This arrangement allowed for the mixing and eventual hybridization of distinct species that were normally separated geographically. Unless a perpetrator claims responsibility, it may be very difficult to determine whether outbreaks of new diseases by previously undescribed pathogens were caused by the trade-facilitated evolution of new pathogen species or the deliberate introduction of novel pathogens created in a laboratory.

In addition to masking acts of bioterrorism and biocrime in the backdrop of frequent accidental introductions, global trade in plants and plant products creates the added problem of providing possible pathways for intentional introductions. There are many access points in global plant distribution networks into which plant pathogens could be easily introduced. Because many of these pathogens are of regulatory concern with quarantine implications, the presence alone of the pathogen could have significant economic impact without actually causing disease in a plant system.

Assigning attribution in the case of an act of terrorism in a plant system may be further complicated by climate change. Among the many impacts predicted by climate change models are climate-induced changes in plant physiology, the geographic redistribution of plant populations, the emergence of new pathogens or new pathogen–vector associations, and the geographic redistribution of existing pathogens (Price-Smith 2002). These issues will make it difficult to determine if an outbreak was the result of an intentional introduction; is a new outbreak the result of bioterrorism or a change in local environmental conditions that allowed the establishment of a pathogen and the development of disease in a previously non-receptive environment?

10.2.2 Specific Threats to Plant Biosecurity

There are several lists of specific pathogens that are potential threats to plant biosecurity (Table 10.1). The different approaches taken to identify threats can explain, in part, the discrepancies among the lists. Most of the organisms identified in these lists are specific with respect to the plant species at risk; for example, pathogen threats to soybean production systems in the United States, pathogen threats to forest ecosystems in Europe. One limitation of these lists of specific threats is our inability to predict invasiveness with any degree of certainty; that is, which species will get introduced into a specific geographic area, and more importantly, which species will get established after an introduction and then, once established, spread beyond the outbreak area. Such lists focus too much attention on the prevention and response to a few pathogens when the systems they are designed to protect are at risk to many pathogens that have a high probability of being introduced accidentally. Equal effort should be placed on developing comprehensive biosecurity strategies for plant systems in addition to specific plans for specific pathogens.

10.3 Bioterrorism as a Threat to Plant Biosecurity

The concept of bioterrorism against a plant system or agricultural production system is difficult for some to accept. After all, inherent in the name and the concept is terror. It is difficult to imagine anyone being terrorized by diseased plants. Leaves with spots, rotting tubers, or even corn plants falling over from stalk rot are not likely to send people running. When diseased plants make the news, it is in reference to the economic or ecological damage incurred; disappointment yes, terror no. The term bioterrorism though is as much about motive as it is about affect. Agroterrorism (or bioterrorism) against plants is conceptually more about why the act might be committed rather than the emotional response of those affected. The most important consideration is the impact from an introduction whether deliberate, accidental, or natural. Bioterrorism is one more threat to consider when developing a strategy for plant biosecurity.

Why would anyone target a plant system? This question goes to motive, which is a function of human thought and behavior. There is no exact answer. Why would anyone write a computer program, attach it to an email, and send that email around the world so that when it is opened a virus infects the hard drive and destroys the computer of someone they never met? This happens every day and has caused immeasurable impact. The only apparent motive is to disrupt. Is bioterrorism against plant systems as significant a risk as global trade in plants? No, it is not. But to not consider it a possible threat is an unnecessary risk.

Table 10.1 Lists of plant pathogens that could potentially be used in acts of agroterrorism around the world (referenced on the Internet in September 2008, see footnotes) (this table is after Table 10.1 in Latxague et al. 2007; with the publisher's permission)

Code	Origin	Fungi	Bacteria	Viruses
BTWC-SA	Ad Hoc Group of the Biological and Toxin Weapons Convention (WP124 by South Africa)[a]	13	6	1
BTWC-AHG	Ad Hoc Group of the Biological and Toxin Weapons Convention (Procedural Report 56/1)[b]	4	3	1
USDA-APS	Animal and Plant Health Inspection Service (APHIS) of the United States Department of Agriculture (USDA), in accordance with the American Phytopathological Society (APS)[c]	4	5	1
USDA-APHIS	Animal and Plant Health Inspection Service (APHIS) of the United States Department of Agriculture (USDA), in accordance with the Agricultural Bioterrorism Protection Act[d]	3	5	0
European Union	EU Plant Health Directive 2000/29/CE[e]	19	3	34
EPPO	Lists of pests recommended for regulation in the European and Mediterranean Plant Protection Organization (EPPO) region[f]			
	A1 List	38	11	23
	A2 List	20	22	19
CNS	Center for Non-proliferation Studies (CNS) at the Monterey Institute of International Studies[g]	18	11	3
AG	Australia Group[h]	8	6	3
ISSG-IUCN	Invasive Species Specialist Group (ISSG), a part of the Species Survival Commission of the World Conservation Union (IUCN)[i]	3	1	0
CBWinfo	Chemical and Biological Weapons Info (CBWinfo) website[j]	27	17	1

[a] http://www.bradford.ac.uk/acad/sbtwc/ahg34wp/wp124.pdf
[b] http://www.bradford.ac.uk/acad/sbtwc/ahg56/doc56-1.pdf
[c] http://www.apsnet.org/members/ppb/RegulatoryAlerts/FEDREG8-12-02.pdf
[d] http://www.biosafety.msu.edu/selectagents/Select_Agent_List.pdf
[e] http://www.boku.ac.at/IAM/pbiotech/eppl.pdf
[f] http://archives.eppo.org/EPPOStandards/PM1_GENERAL/pm1-02(16)_A1A2_2007.pdf
[g] http://cns.miis.edu/research/cbw/biosec/pdfs/agents.pdf
[h] http://www.australiagroup.net/en/plants.html
[i] http://www.issg.org/booklet.pdf
[j] vhttp://www.cbwinfo.com/Biological/PlantPath.html

10.3.1 Is Bioterrorism a Real Threat to Plant Biosecurity?

Bioterrorism targeting agricultural systems has been identified as a significant concern by many and of questionable significance by some (Wheelis et al. 2002; Cupp et al. 2004; Nutter and Madden 2005; Young et al. 2008; Wheelis et al. 2008). In light of the potential to influence international travel and trade policies, the funding of research and education programs, and the development of national preparedness plans, it is appropriate to question whether bioterrorism is a real threat to plant biosecurity. Often, the discussion of plant biosecurity and agro-terrorism is evaluated from the perspective human systems. However, the nature and the magnitude of value in plant systems are inherently different requiring a different approach to risk assessment. For plant systems, the intent may be to reduce food production capacity, to render food unpalatable/harmful, to undermine public confidence in food production and food safety systems, and to cause large-scale and sustained economic damage that ultimately lowers a nation's standard of living. To lower a developed nation's food production capacity and/or standard of living would likely take a long period of time and be very difficult to accomplish. Most developed nations have adequate to excess capacity to produce food and/or multiple trade agreements to compensate for deficiencies. To lower a developing nation's food production capacity and/or standard of living could be accomplished in a very short period of time and with relative ease. Many developing nations lack the capacity to produce adequate food and are resource-poor precluding trade to compensate for deficiencies. These nations are very vulnerable to agro-terrorism.

To render food unpalatable/harmful or to inflict significant economic damage to food production systems could be accomplished in a very short period of time in any nation. Many agricultural production and distribution systems are open systems with many possible pathways for the intentional introduction of pathogens. Depending upon the pathogen and the plant system targeted, it would be possible to inflict significant economic damage without causing an epidemic. A quarantine pest or pathogen need only be detected to stop shipment of plants or plant products; it does not actually have to cause disease. Another possible objective might be to destabilize international relations by causing sustained disruption of trade among signatories to bilateral and multilateral trade agreements. Recent shortages in rice and wheat supplies resulted in bans of exports by several nations (IFPRI 2005; Shelburne 2008). Disagreements over trade can create tension between nations. The intentional introduction of a pathogen to reduce the production capacity upon which a trade agreement is based, or a quarantine pathogen to disrupt a trade agreement, could strain international relations and strategic alliances.

Over the last 50 years, most acts of terrorism have targeted humans and/or infrastructure with explosives and incendiary devices. Bombs have immediate and dramatic impacts that provide gripping visual images that extend the impact beyond the targeted area. Two of the most commonly cited objectives of terrorists are to instil fear and to disrupt socio-political systems. Clearly, bombs have those effects. During the previous 30 years, there were several acts of bioterrorism against humans, including, the unsuccessful anthrax attack in Japan and the successful anthrax and

Salmonella attacks in the United States (Ostfield 2007). The anthrax events in the U.S. caused fear in the general population and disruption of socio-political systems as evidenced by the interruption of sessions of the U.S. Congress and the local suspension of normal operations of the U.S. postal system. Would the introduction of a plant pathogen or an insect pest with the ability to cause disease or damage to a natural or agricultural plant system have immediate and dramatic effects? Probably not, but one must consider the spatial and temporal scales of impacts. For human and infrastructure targets, the scale can be quite small yet result in very large impacts. One to several human deaths or one large building damaged can cause fear and disruption in both local and regional populations; the threat is overt and the impact direct. The number of humans directly affected by the U.S. anthrax letter attacks was 27, including five deaths. However, 300 million people were indirectly impacted to varying degrees. The destruction of two buildings (i.e., the World Trade Center Towers) in New York in 2001 had enormous direct local impacts and long lasting global impacts. Fear, disruption, and immeasurable economic impacts resulted from those two terrorist actions. The long-term effects included costly and cumbersome security measures in the form of building infrastructure enhancements, stringent policies regarding human behavior, and the implementation of a legal system that governs the possession, use and transportation of certain microorganisms. Equally costly and cumbersome security measures impacted the transportation sector as well (e.g., infrastructure enhancements, stringent human behavior policies, international travel agreements, strict passport policies). It has been several years since those bioterrorism events occurred, yet the impacts are still evident.

One to several plant deaths, if even noticed, will have little socio-political effect unless those plants are of cultural importance (e.g., the Treaty Oak in Texas that has deep historical and cultural significance, Maraniss 1989), or of deep personal importance (e.g., citrus trees on individual homeowners property, Gottwald et al. 2002). The value in plants and plant systems is usually found in populations of plants or the commodities derived from those plants, not in the individual plants. Consequently, the appropriate concern with plant systems should be measured over time and over large spatial scales. It is possible that acts of bioterrorism targeting plant systems could elicit large-scale effects over time with limited prospects for complete recovery.

10.3.2 Evidence for Bioterrorism in Plant Systems

Historical Perspective Several nations developed the technologies necessary for the large-scale production and deployment of plant pathogens (Rogers et al. 1999; Suffert 2003; Madden and Wheelis 2003). Whether these bioweapons were ever used overtly or clandestinely to target plant systems is not known. Small-scale field tests were successful in demonstrating their effectiveness. Proof of concept has been established.

The Brazilian Cocoa Case Approximately 20 years ago, the intentional introduction of the witches broom pathogen (*Crinipellis perniciosa* (Stahel) Singer) into cocoa plantations in the Bahia region of Brazil was alleged by cocoa producers in the affected area

(Homewood 1991, Junior 2006). Cocoa branches with disease symptoms were reported to have been found wired to trees at the outbreak site. Epidemiologists concluded that the natural dispersal of *C. perniciosa* spores from the Amazon cocoa production area to the Bahia production area was unlikely. Land reform activists were blamed for the act in an attempt to destabilize the local government. Thousands of trees were ultimately affected reducing cocoa yields by 75% and causing serious economic losses (Bowers et al. 2001). Years later, a man confessed to the act of deliberately introducing the pathogen in order to undermine the local government (Junior 2006).

Unsubstantiated Cases There have been many cases of one nation accusing another of using biological weapons against agricultural plant systems (Junior 2006; Suffert et al. 2008; Zilinskas 1999). There is little, if any, compelling evidence to support these claims. The general lack of forensic technologies and protocols specific to plant pathosystems coupled with the existence of natural pathways for the introduction of the plant pathogens associated with these cases make difficult establishing that an introduction was deliberate. Assigning attribution would be extremely difficult unless those responsible claimed responsibility. However, the inability to prove culpability does not equate to proof of innocence.

10.3.3 Plant Systems as Soft Targets

Agricultural systems are vulnerable because of their economic and sociologic importance. Crop and forest systems are vulnerable because they are grown on large, unsecured, poorly monitored areas. Historical evidence indicates that agroterrorism (i.e., anticrop bioterrorism, biowarfare, and biocrime) is not just an academic issue. Throughout history, agricultural systems have been targets in war; crops and forests were trashed or burned to deprive the enemy of food thereby repelling colonists or subjugating rebel populations. During and after the Second World War, several countries developed research programs for biological anti-crop agents targeting the worlds staple crops (*Phytophthora infestans*, agent of potato late blight, *Cochliobolus miyabeanus*, agent of rice brown spot, and *Magnaporthe grisea* agent of rice blast) (Foxwell 2001; Suffert 2002, 2003; Madden and Wheelis 2003). After the Cold War, several countries continued to conduct research on plant pathogens as anti-crop weapons, including *Puccinia graminis* f. sp. *tritici* the causal agent of wheat stem rust (Line and Griffith 2001; Whitby 2002). While countries that signed the Biological and Toxin Weapons Convention (BTWC) in 1972 officially stopped their biological warfare programs, a new cycle of concern over the possible use of biological anti-crop weapons began in the late 1980s, based on the knowledge that several "rogue" countries were developing such weapons (e.g. the wheat smut fungi *Tilletia caries* and *T. tritici* in Iraq). Additionally, there have been sporadic allegations that states have either used plant pathogens against crops or threatened to use them for political purposes. Cuban authorities alleged without significant evidence, that the introduction in the 1970s of *Personospora hyoscyami* f. sp. *tabacina*, the causal agent of tobacco blue mold, and *Puccinia melanocephala*, the causal agent of sugarcane rust, were the results of anti-crop attacks by the US

(Zilinskas 1999). Recent evidence found in caves in Afghanistan suggested interest by Islamic militants in the weaponization of wheat rust (Fletcher et al. 2006). In the 1990s, the United Nations Drug Control Program sponsored anti-coca (using *Fusarium oxysporum* strains) and anti-poppy (using *Pleospora papaveracea* strains) research programs in Andean and Central Asian countries, respectively (Connick et al. 1998; O'Neill et al. 2000); these were officially never used. The use of biocontrol agents as biowarfare agents is controversial (Suffert et al. 2008). From a scientific point of view, the relevance of these drug-control programs to agroterrorism is that the methods used are the same as those used in state-sponsored programs, including the preparation and storage of large amounts of inoculum and the delivery of inoculum clandestinely for the purpose of destroying a cultivated crop.

10.3.4 The Technology Factor

The state-sponsored bioweapons programs of the past were not limited by lack of infrastructure and funding; long-term research programs and large-scale production systems were possible. However, advances in the technologies that underpin modern biological sciences, communication systems and transportation systems overcome the need for large scale infrastructure and funding. The biotechnology industry has made amazing advances in microbe and cell technologies; for example, genetic manipulation, fermentation, and stabilization. Communications technologies and systems (e.g., Internet) make research results and protocols available to everyone worldwide in near real time. Transportation systems move materials around the world in very short periods of time. It is conceivable that someone could set up a laboratory in a concealed location, engineer a plant pathogen with novel virulence traits, produce a significant quantity of the novel pathogen in a stable formulation, and ship that pathogen around the world without being discovered. The technologies that improve our lives also increase our vulnerabilities. Scientific advancements that can be misused have been termed dual use technology. Dual use technology has been the subject of much discussion over the need for regulation of such technology. The dual use dilemma is defining the characteristics that make research dual use and developing an appropriate system of regulations that enhance safety without compromising scientific progress (IOM/NRC 2006). The risk of not doing this type of research may be much greater than the risk posed by the potential for misuse.

10.4 Threat and Vulnerability Assessments

10.4.1 Strategy

There have been very few attempts to propose a methodology specific to the assessment of agroterrorism, which has often been described as "low-tech, high impact" requiring "relatively little specialized expertise and technology" (Rogers et al., 1999;

Wheelis et al. 2002). Most of these discussions are not based on a quantitative analysis of the threat. The success of a malevolent act might be much more uncertain than believed.

Madden and van den Bosch (2002) designed a probabilistic method to compute a global risk index for a plant pathogen, based on the product of the probabilities of single events required for a successful agroterrorist attack (i.e., pathogen introduction, disease establishment, spread, damage and lack of control measures). Extremely low values for some intermediate probabilities and the lack of validation of the assessment method limit the utility of the method for real cases. Schaad et al. (2006) applied an analytic hierarchy process, based on a set of prioritized and qualitatively assessed (high, mean, low) criteria, to eight potato pathogens. Both approaches synthesized the information obtained from a panel of experts in a single risk value rather than in a risk profile. Despite the involvement of pathogen experts, these probabilistic methods appear more useful as theoretical exercises than as tools for stakeholders. In such approaches, the risk may be overestimated for a plant pathogen well-known to many experts in the assessment group.

A third approach, based on the perpetrators' presumed objectives and the expected consequences, emphasized the direct economic consequences of the act; for example, crop loss (Latxague et al. 2007; Suffert et al. 2008). It included psychological and indirect economical (e.g., trade disruption, penury) consequences, which are presumed objectives of an agroterrorist attack (Huff et al. 2004; Waage and Mumford 2007).

In the absence of an unambiguous definition of agroterrorism, a subject of debate among plant pathologists, Suffert et al. (2008) proposed that the risk should be characterized by a foresight approach, which took into account the hybrid nature of the threat, the multiplicity of the perpetrator's objectives and expected consequences and the diversity of *modus operandi*. The methodology includes three successive steps (Fig. 10.1):

1. Build a list of 50 candidate plant pathogens representing potential threats to European agriculture and forest systems (Latxague et al. 2007).
2. Develop a scenario-based, foresight investigation of potential agroterrorist acts in Europe and assign a key pathogen from the candidate list to each of the nine proposed scenarios. Three types of acts were considered (international, state-sponsored biowarfare; non-governmental bioterrorism; and individual or corporate biocrime). Combining the nature of the acts and their potential consequences, nine scenarios of agroterrorist attacks, divided into three sections (synopsis, justification, feasibility), were developed and their socio-economical consequences investigated (Latxague et al. 2007).
3. Design a risk assessment scheme (RES), derived from a standard Pest Risk Analysis (PRA) (IPPC 2004; EPPO 2007) and apply it to key pathogens. The RES included five sections (importance of the target crop; ease of use of the pathogen; epidemic potential of the pathogen; obstacles to swift and effective response to an attack; and potential global or regional consequences of an attack) scored using criteria documented with scientific literature (Latxague et al. 2007).

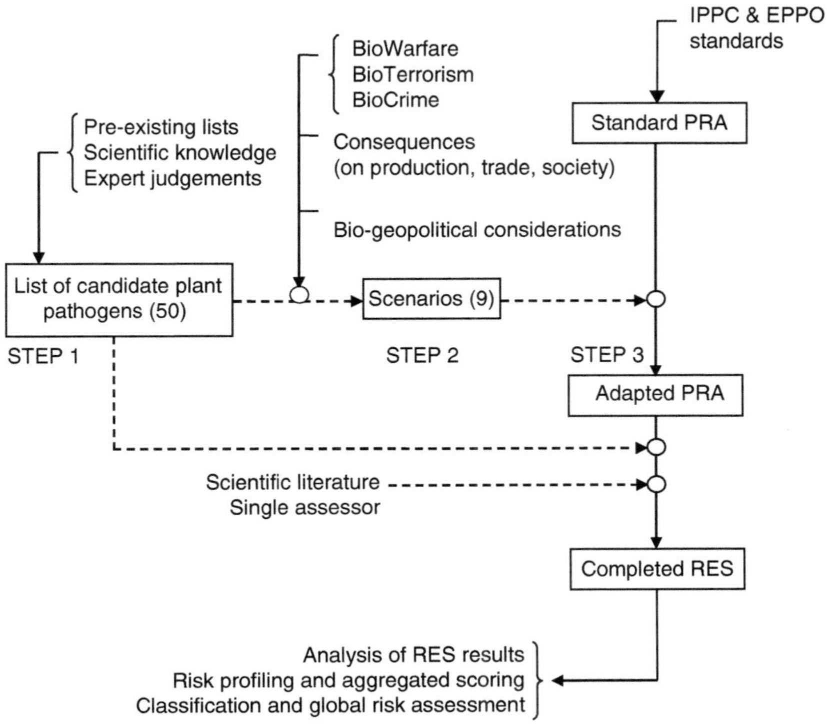

Fig. 10.1 Methodology used by Latxague et al. (2007) for assessing the risk posed by agroterrorism in Europe (after Figure 10.1 in Latxague et al. 2007; with the publisher's permission)

A resulting pentagonal star plot represented the risk profile of each pathogen and an aggregated risk was calculated. This step can be applied by non-experts on particular diseases and thus permits a comparison between crops or pathogens on the basis of the characterization of the threat, and the expected effects of the attack.

10.4.2 Threat Identification and Assessment

Various international working groups and organizations (e.g., the Biological and Toxin Weapons Convention, the Animal and Plant Health Inspection Service of the United States Department of Agriculture, the American Phytopathological Society, the European Union, the European and Mediterranean Plant Protection Organization and the Centre for Nonproliferation Studies) have compiled lists of pathogens of quarantine or agroterrorism concern. Derived list of pathogens were compiled or assessed regarding a specific crop (e.g., potato, Schaad et al., 2006), a pathogen group (e.g., bacteriology; Young et al. 2008), or a target country (e.g., United States; Madden and Wheelis 2003; Slovenia, Boben et al. 2008). Available lists were also critically screened by the partners and experts of the EU CROPBIOTERROR project and updated with relevant scientific information (Gullino et al. 2007). The list of 50 pathogens (35 fungi and oomycetes, nine bacteria and phytoplasmas, and six viruses) included exotic and quarantine pathogens that may induce epidemics causing damage in Europe (e.g., *Ceratocystis fagacearum*, *Erwinia amylovora*, *Mycosphaerella populorum*, *Pepino mosaic potexvirus*, *Pleospora papaveracea*, *Plum pox potyvirus*, *Phakopsora pachyrhizi*, *Ralstonia solanacearum*, *Synchytrium endobioticum*, *Tilletia indica*, *Xylella fastidiosa*), as well as more common indigenous pathogens causing recurrent epidemics. The indigenous pathogens were selected because they represented particular risk profiles; for example, the production of mycotoxins (e.g., *Claviceps purpurea*, *Fusarium graminearum*, *Gibberella zeae*, *Penicillium expansum*) or the existence of exotic strains that could replace or hybridize with local strains (e.g., *Leptosphaeria maculans*, *Phytophthora infestans*, *Puccinia triticina*) (Latxague et al. 2007). Staple food crops represented the majority of targets (24), followed by forest trees (11), industrial and market crops (10), and orchard trees (5). In 32 out of the 50 cases, direct crop loss was predicted following an attack, while trade would be disrupted in 38 out of the 50 cases. Wider, indirect socio-economical consequences, such as poisoning of animals and humans, patrimonial and environmental loss or psychological negative effect on populations, were predicted in 28 cases.

The desired effect of the attack will largely determine the perpetrator's strategy, the target crop, and the pathogen weapon. For example, pathogens with low effect on crop yield could be used for agroterrorism, provided that they disrupt trade via quarantine establishment or produce toxins threatening to human and animal health. A significant practical problem for the perpetrator is to gather the scientific and technical information required for a successful act. The necessary steps in the acquisition of knowledge can make improbable a bioterrorist attack (Suffert et al. 2008). State-sponsored biowarfare scenarios and corporate biocrime scenarios are not resource-limited. Individuals or terrorist groups may require the cooperation of disaffected scientists (a certain scenario has a depraved scientist as the perpetrator), the "phishing" of information in an indirect way and access to laboratory and field facilities. The effective response and management of an agroterrorist attack will depend on the early detection of the pathogen and the rapid implementation of countermeasures; this is similar to the case for the eradication of quarantine pathogens. A state-sponsored biowarfare attack may operate on a large scale with several inoculation sites or utilize an innundative approach in order to overwhelm the countermeasure system set up by the target country. Biocrime operating on a smaller scale may elude early detection thus jeopardizing immediate eradication of the pathogen. In contrast, bioterrorist attacks may be "advertised" by the perpetrators in order to increase psychological confusion and disorganise the countermeasure system; such acts may target crops for which protection measures are difficult to implement or are inefficient.

The aggregated risk score, based upon the sum of risk components, predicted that the importance of the target crop was maximal for wheat and maize and minimal for soybean and poppy; the ease of pathogen use was higher for the saprotrophic pathogens (based on growth on artificial medium) than for the biotrophic pathogens (in most cases not cultivable on artificial medium); the epidemiological potential was maximal for airborne pathogens already established in Europe and adapted to the local environment; the obstacles to swift and effective response to an attack were greatest for pathogens not present in Europe yet and for which quick detection methods are unavailable; and the potential global or regional consequences of an attack was maximal for regulated pathogens (Suffert et al. 2008).

10.4.3 Vulnerability Identification and Assessment

There is broad consensus that the threats to plant biosecurity are increasing due to growing trade, travel, transportation and tourism, the 'four T's' of globalization (Waage and Mumford 2007). A recent scientometric analysis confirmed that the concept of agroterrorism emerged in the scientific literature after 1997 (Suffert et al. 2008); its importance increased after the 11 September 2001 terrorist attacks in the U.S.

Mass destruction of food crops by the introduction of an exotic plant pathogen seems highly improbable in most advanced industrialised countries. However, the malevolent use of plant pathogens could have high social and economic impacts. Moreover, the biosecurity of forest and natural ecosystems should be considered as a serious issue (Cochrane and Haslett 2002). Plant pathogenic fungi that produce mycotoxins, with the potential to affect human and animal health should be considered legitimate threats even though most of them are already a recurrent cause of disease. The low production of mycotoxins and the availability of adequate detection methods lead some to question mycotoxin-producing fungi as serious anti-crop agents (Paterson 2006). However, a deliberate introduction of a toxin-producing plant pathogen could cause significant disruption and a loss of confidence in the food chain. Additionally, a perpetrator with limited technical and scientific skills, using simple intimidation or blackmail could circumvent the unpredictable success of a deliberate contamination by issuing false claims (Huff et al. 2004; Waage and Mumford 2007). These types of events could strain international relations with global consequences (Castonguay 2005; Hennessy 2008).

In two third of the scenarios proposed by Suffert et al. (2008), the perpetrators would conceal their action, while in one third of the scenarios they would claim responsibility. To determine the source of the pathogen, the methods employed, the time of the introduction, and the identity of the perpetrators would require a forensic investigation (i.e., the application of scientific methods in the investigation of possible violations of the law, where scientific knowledge and technology provide evidence in both criminal and civil matters) (Fletcher et al. 2006). Early detection and identification of plant pathogens would be critical to a forensic investigation and the determination that an epidemic was of "suspicious" origin (Stack et al. 2006; Waage and Mumford 2007; Suffert et al. 2008). The longer an outbreak was developing and progressing into an epidemic, the more difficult it would be to detect signal in the background of noise. Those attempting to gather evidence that might aid in determining cause, whether intentional or accidental, will be challenged as the size of the impacted area and the severity of disease increases. This will be true whether the perpetrator claims responsibility or not.

In the context of the World Trade Organisation (WTO) and the Sanitary and Phytosanitary Agreement (SPS), it is possible that a country, in bad faith, could use "agroterrorism" as a justification to impose trade barriers. It is also conceivable that a "rogue" country would clandestinely introduce a regulated pathogen into a shipment imported from another country to justify protecting its traditional markets. The international context makes plausible the bio-geopolitical justification of some biowarfare scenarios (Wheelis et al. 2008; Suffert et al. 2008).

10.5 Security Through International Cooperation

Functionally, the world is a smaller place today. Rapid intercontinental flights speed travelers from hemisphere to hemisphere and high-speed trains from one side of a continent to the other in hours. Demographics are changing as immigrants from politically unstable, war-ravaged or poverty-plagued nations seek better lives in other countries. International commerce is booming; the most-exchanged commodities include thousands of agricultural products. Agriculture and the food chains that support humanity are vulnerable targets; an attack on them could have devastating consequences, not just for health and safety, but also in terms of social and economic impact. It is extremely important to increase international cooperation on biodefense to protect agriculture and food systems worldwide. International cooperation is needed at the scientific, policy, legal, and commercial levels permitting the sharing of views and information. Convergence of ideas and experiences is needed to enhance global preparedness. International cooperation is critical to understanding and addressing the effects of implementing new or enhanced food defense measures on various components of agro-food industries. Agriculture and the food production and distribution systems we depend upon are global in scope requiring an international approach to plant biosecurity (Gullino et al. 2008).

10.5.1 Science

Scientific cooperation often provides a sound basis for cooperation among countries. Scientific communication is different from diplomatic communication; it is based upon the free exchange of ideas and emphasizes commonalities rather than differences. Scientists are outstanding ambassadors for their nations; they interact at a peer level and share mutual respect for their science. The basic needs for knowledge and technology are common to all nations; what varies is the ability of each nation to develop, acquire, and implement new technologies. Duplication of effort in the development, screening, standardization and validation of molecular or serological tests, is inefficient and costly. Networking researchers and plant protection practitioners has many advantages, including enhanced diagnostic, communication and training capabilities. Cooperation will lead to the development and sharing of new standards for diagnostic procedures, for the validation of such procedures and for laboratory certification. International cooperation will accelerate the development of standards and protocols by which to differentiate plant pathogens at the subspecies or isolate level, to identify genetic modifications of known agents, and to permit the determination of events as intentional, accidental or naturally occurring. Collaboration in the development and validation of pathogen modeling and risk analysis tools will enhance security and permit the integration of epidemiology and economic risk assessment into policy formulation. Beyond traditional detection and diagnostics, the new discipline of forensic microbiology (including forensic plant pathology) is another area in which international collaboration will be beneficial. The sharing of sequence data and microbe culture collections will be important to rapid advancement in this emerging discipline. The ability to share information instantly and globally through informational websites, interactive online chat rooms, linked home pages, teaching aids, directories, image libraries and innovative e-publications is limited only by creativity and resourcefulness.

Productive international collaborations can be variable in the number of countries participating as well as in the basis of cooperation: for example, common vulnerabilities and threats, complementation or synergy in technological capabilities and scientist training/experience, etc. It may also be based upon geography (countries of a single region, e.g., European Union) or trade agreements (e.g., the United States and the EU). The goal in any case should be to create mechanisms for exchange of information on diseases of concern. Some European research networks are already addressing issues related to bio-weapons. Among them are Consortia on crop biosecurity (CROP BIOTERROR, TOOLS FOR CROP BIOSECURITY and BIOSEC); WATERSAFE which focuses on drinking water, AEROBACTIS on airborne microorganisms; ASSRBCVUL on radiological, biological and/or chemical agents; BIOSAFENET on genetically modified organisms; and EPIZONE on epizootic diseases in agriculture and aquaculture.

10.5.2 Policy

Officials of governments generally communicate through formal and informal channels. Because they represent sovereign states, they provide official viewpoints and approved positions. The formality of such communication is necessary because of the authority and impact of those interactions. Common policies are important to protect global agriculture and food. In 2004, bioterrorism was included on the G8 Agenda, leading to a statement regarding the issue of "Defending against bioterrorism". In 2005, G8 nations built on this policy foundation and established some of the first-ever international technical and policy initiatives for food defense.

The Asia Pacific Economic Cooperation (APEC) forum also addresses food defense (Ostfield 2007). It is critical that scientists participate in the policy making process by providing the information necessary to ensure that rational plant biodefense policies, based upon sound scientific data, are formulated and adopted.

10.5.3 Law

Preparedness for, prevention of, and response to acts of bioterrorism and biocrime directed against agriculture and food will require international cooperation. Such acts involve breaking one or more national and/or international laws. One specific area for international cooperation is the development and application of microbial forensic technologies and protocols to ensure that attribution is successful and that justice is served following the intentional introduction of a plant pathogen across national borders (Fletcher 2008).

10.5.4 Commerce

Implementing new or enhanced biodefense measures for regulation and oversight might seriously impact public and private components of the agricultural and food industries, particularly small and medium enterprises, ultimately affecting global trade in food and agricultural products. In the event of a terrorist attack, international cooperation will be challenging, with the potential to create short term tension among trade partners, and potential long term and lasting diplomatic tensions (Ostfield 2007).

It is extremely important to increase international cooperation on biodefense to protect agricultural and food systems worldwide. Promoting convergence of ideas and experiences should be a key element of cooperation among all nations, in particular those with trade agreements involving food and other agricultural products. Collaborations will enhance global preparedness while creating mechanisms for understanding the impacts of food defense policies on the agro-food industries within each nation.

Acknowledgement This work was funded by a grant from the European Commission for the project "Crop and food biosecurity, and provision of the means to anticipate and tackle crop bioterrorism", Contract n. 6403 and by a grant from _______________ to the National Agricultural Biosecurity Center for the project __________ Contract n. ______.

References

Boben J, Urek G, Radisek S, Mehle N, Dreo T, Širca S, Pirc M, Žerjav M, Virš ek-Marn M, Ravnikar M (2008) Crop biosecurity in Slovenia: strategies and implementation. J Plant Pathol 90(2):148

Bowers JH, Bailey BA, Hebbar PK, Sanogo S, Lumsden RD (2001) The impact of plant diseases on world chocolate production. Online. Plant Health Prog . doi:10.1094/PHP-2001-0709-01-RV
Brasier CM (2001) Rapid evolution of introduced plant pathogens via interspecific hybridization. Bioscience 51(2):123–133
Brasier CM (2008) The biosecurity threat to the UK and global environment from international trade in plants. Plant Pathol 57(5):792–808
Britton KO (2004) Controlling biological pollution. pp 1–7. In: Kerry Britton (ed) Biological pollution, an emerging global menace. American Phytopathological Society, St. Paul, MN, 113 pp
Castonguay S (2005) Biorégionalisme, commerce agricole et propagation des insectes nuisibles et des maladies végétales: les conventions internationales phytopathologiques. Ruralia 16(17):137–52
Cochrane H, Haslett D (2002) Deliberate release – what are the risks? NZ J Forestry 47:16–7
Connick WJ, Daigle DJ, Pepperman AB, Hebbar KP, Lumsden RD, Anderson TW, Sands DC (1998) Preparation of stable, granular formulations containing *Fusarium oxysporum* pathogenic to narcotic plants. Biol Control 13:79–84
Cupp OS, Walker DE, Hillison J (2004) Agroterrorism in the US: Key security challenge for the 21st century. Biosecur Bioterror 2:97–105
EPPO (2007) PM 5/3 (3) Decision-support scheme for quarantine pests [http://www.eppo.org/QUARANTINE/Pest_Risk_Analysis/PRA_intro.htm].
Fletcher J (2008) The need for forensic tools in a balanced national agricultural security programme. In: Gullino ML, Fletcher J, Gamliel A, Stack JP (eds) Crop biosecurity: assuring our global food supply. Springer, The Netherlands, 148 pp
Fletcher J, Bender C, Budowle B, Cobb WT, Gold SE, Ishimaru CA, Luster D, Melcher U, Murch R, Scherm H, Seem RC, Sherwood JL, Sobral BW, Tolin SA (2006) Plant pathogen forensics: capabilities, needs, and recommendations. Microbiol Mol Biol Rev 70:450–71
Foxwell J (2001) Current trends in agroterrorism (antilivestock, anticrop, and antisoil bioagricultural terrorism) and their potential impact on food security. Stud Conf Terror 24:107–29
Gottwald TR, Graham JH, Schubert TS (2002) Citrus canker: The pathogen and its impact. Online. Plant Health Prog . doi:10.1094/PHP-2002-0812-01-RV
Gullino ML, Suffert F, Dehne H, Thomas J, Barker I, Gamliel A, Bonifert M, Stack J, Fletcher J, Abd-Elsalam K (2007) Crop and food biosecurity: first results of European research. Phytopathology 97:S44
Gullino ML, Fletcher J, Gamliel A, Stack J (eds) (2008). Crop biosecurity: assuring our global food supply. Springer, The Netherlands. 148 pp
Hennessy DA (2008) Economic aspects of agricultural and food biosecurity. Biosecur Bioterror 6:66–77
Homewood B (1991) Fungus threat to Brazil's 300 million cocoa trees. New Scientist 1778:16
Huff KM, Meilke KD, Turvey CG, Cranfield J (2004) Modeling bioterrorism in the livestock sectors of NAFTA members. Curr Agric Food Resour Issues 5:1–22
IOM/NRC (2006) Globalization, biosecurity, and the future of the life sciences. Institute of Medicine and National Research Council Report. The National Academies Press, Washington DC, 299 pp
IPPC (2004) Pest risk analysis for quarantine pests, including analysis of environmental risks and living modified organisms. International Standard for Phytosanitary Measures no. 11, FAO, Roma, Italy
Junior P (2006) Terrorismo biologico. Veja 1961:60–63
Latxague E, Sache I, Pinon J, Andrivon D, Barbier M, Suffert F (2007) A methodology for assessing the risk posed by the deliberate and harmful use of plant pathogens in Europe. EPPO Bull 37:427–35
Line RF, Griffith CS (2001) Research on the epidemiology of stem rust of wheat during the Cold War. In: Peterson PD (ed) Stem rust of wheat: from ancient enemy to modern foe. APS Press, St. Paul, US, pp 83–118
Mack RN, Lonsdale WM (2001) Humans as global plant dispersers: getting more than we bargained for. Bioscience 51(2):95–102
Madden LV, Van Den Bosch F (2002) A population-dynamics approach to assess the threat of plant pathogens as biological weapons against annual crops. BioScience 52:65–74
Madden LV, Wheelis M (2003) The threat of plant pathogens as weapons against US crops. Annu Rev Phytopathol 41:155–76

Man in't Veldt WA, Veenbaas-Rijks WJ, Ilieva E, de Cock AWAM, Bonants PJM, Pieters R (1998) Natural hybrids of *Phytophthora nicotianae* and *Phytophthora cactorum* demonstrated by isozyme analysis and random amplified polymorphic DNA. Phytopathology 88:922–929
Maraniss D (1989) Texas mourns imminent death of 500-year old Treaty Oak; Austin landmark apparently was poisoned. Washington Post, June 27, 1989
Nutter FW Jr, Madden LV (2005) Plant diseases as a possible consequence of biological attacks. In: Greenfield RA, Bronze MS (eds) Biological terrorism. Horizon Scientific Press, Caister Scientific Press, Norfolk, UK, pp 793–818
O'Neill NR, Jennings JC, Bailey BA, Farr DF (2000) *Dendryphion penicillatum* and *Pleospora papaveracea*, destructive seedborne pathogens and potential mycoherbicides for *Papaver somniferum*. Phytopathology 90:691–8
Ostfield ML (2007) Biodefense: U.S. vision of a broader cooperation. Eur Aff 8(1):75–79
Palm M (1999) Mycology and world trade: a view from the front line. Mycologia 91:1–12
Paterson RRM (2006) Fungi and fungal toxins as weapons. Mycol Res 110:1003–10
Price-Smith A (2002) The health of nations. MIT Press. Cambridge, MA, 220 pp
Rogers P, Whitby S, Dando M (1999) Biological warfare against crops. Sci Am 280:70–5
Rossman A (2001) A special issue on global movement of invasive plants and fungi. BioScience 51(2):93–94
Schaad NW, Abrams J, Madden LV, Frederick RD, Luster DG, Damsteegt VD, Vidaver AK (2006) An assessment model for rating high-threat crop pathogens. Phytopathology 96:616–21
Shelburne EC (2008) The great disruption. The Atlantic Monthly, September, pp 28–29
Stack JP (2008) Challenges to crop biosecurity, pp 15–23. In: Gullino ML, Fletcher J, Gamliel A, Stack PJ (eds) Crop biosecurity: assuring our global food supply. Springer, The Netherlands, 148 pp
Stack JP, Fletcher, J (2007) Plant biosecurity infrastructure for disease surveillance and diagnostics. pp 95–106. In: Institute of Medicine. 2007. Global infectious disease surveillance and detection: assessing the challenges – finding the solutions. The National Academies Press, Washington, DC, 263 pp
Stack J, Cardwell K, Hammerschmidt R, Byrne J, Loria R, Snover-Clift K, Baldwin W, Wisler G, Beck H, Bostock R, Thomas C, Luke E (2006) The national plant diagnostic network. Plant Dis 90:128–36
Suffert F (2002) L'épidémiologie végétale, nouvelle discipline de guerre? Lumière sur le bioterrorisme agricole, un enjeu émergent pour la recherche agronomique. Le Courrier de l'Environ 47:57–69
Suffert F (2003) L'utilisation volontaire d'agents phytopathogènes contre les cultures. L'agroterrorisme et ses conséquences sur notre approche de la lutte contre les maladies des plantes. Phytoma 563:8–12
Suffert F, Barbier M, Sache I, Latxague E (2008) Biosécurité des cultures et agroterrorisme : une menace, des questions scientifiques et une réelle opportunité de réactiver un dispositif d'épidémiovigilance. Le Courrier de l'Environ
Waage JK, Mumford JD (2007) Agricultural biosecurity. Philos Trans R Soc B 363:863–76
Wheelis M, Casagrande R, Madden LV (2002) Biological attack on agriculture: low-tech, high-impact bioterrorism. Bioscience 52:569–76
Wheelis M, Yokoyama V, Ramos C (2008) Agricultural warfare and bioterrorism using invasive species. In: Heather NW, Hallman GJ (eds) Pest management and phytosanitary trade barriers. CABI, Wallingford, United Kingdom, pp 14–19
Whitby S (2002) Biological warfare against crops. Palgrave, Basingstoke, United Kingdom
Windle PN (2004) Exotic pests: past, present, and future, pp 17–27. In: Kerry Britton (ed) Biological pollution, an emerging global menace. APS Press, St. Paul, MN, 113 pp
Young JM, Allen C, Coutinho T, Denny T, Elphinstone J, Fegan M, Gillings M, Gottwald TR, Graham JH, Iacobellis NS, Janse JD, Jacques MA, Lopez MM, Morris CE, Parkinson N, Prior P, Pruvost O, Rodrigues Neto J, Scortichini M, Takikawa Y, Upper CD (2008) Plant-pathogenic bacteria as biological weapons – real threats? Phytopathology 98(10):1060–1065
Zilinskas RA (1999) Cuban allegations of biological warfare by the United States: assessing the evidence. Crit Rev Microbiol 25:173–227

Chapter 11
The Revised International Plant Protection Convention – a New Context for Plant Quarantine

William Roberts

The international agreement covering plant pests and diseases is the International Plant Protection Convention (IPPC). The first version of this convention was adopted in 1952 with a revised version coming into force in 2005 (FAO 1999). Currently 170 countries are contracting parties to the IPPC.

As set out in the text the purpose of the convention is: "securing common and effective action to prevent the spread and introduction of pests of plants and plant products and to promote appropriate measures for their control" (Article 1). The Convention uses the term "pest" to include "any species, strain or biotype of plant, animal or pathogenic agent injurious to plants or plant products". The term plants and plant products covers living plants, material of plant origin but also manufactured products that by their nature or their processing may create a risk of the introduction and spread of pests. These definitions give the Convention a very wide scope covering all situations where there are risks of spread of plant pests "particularly where international transportation is involved" (Article 4).

The adoption of the WTO Sanitary and Phytosanitary Agreement (SPS) in 1995 emphasised the important role of the IPPC by recognizing the IPPC as the appropriate international standard setting organization relevant to plant pests. It was this recognition that provided the impetus to revise the Convention. This revision updated the Convention language and established the Commission on Phytosanitary Measures (CPM) to direct the activities of the Convention. The CPM is the meeting of the countries that are members of the Convention. It meets annually at the United Nations Food and Agriculture Organization in Rome, Italy.

Although recognised by the SPS Agreement the IPPC is not simply limited to trade, it is relevant to agricultural and environmental protection generally. Its scope partly overlaps with the scope of the Convention on Biological Diversity (CBD) and the Secretariats of both the IPPC and the CBD work cooperatively in areas of mutual interest.

W. Roberts (✉)
Biosecurity Australia, GPO Box 858, Canberra, ACT, 2601, Australia
e-mail: bill.roberts@biosecurity.gov.au

R.N. Strange and M.L. Gullino (eds.), *The Role of Plant Pathology in Food Safety and Food Security,* Plant Pathology in 21st Century, 3, DOI 10.1007/978-1-4020-8932-9_11,

The main activity of the IPPC is the development of international standards. These are called International Standards for Phytosanitary Measures (ISPM). These provide guidance to IPPC members on the implementation of the rights and obligations of the convention. Currently there are over 30 ISPMs (www.ippc.int), covering topics such as:

- Principles of plant quarantine as related to international trade
- Risk analysis
- Release of biological control agents
- Establishment of pest free areas or areas of low pest prevalence
- Treatments to kill or remove pests
- Use of systems approaches to pest risk management
- Regulatory and certification systems
- Inspection and sampling methodology

Initially the ISPMs developed dealt with high level concepts but increasingly specific ISPMs dealing with individual pests and specific diagnostic tests are being developed.

One of the major challenges in preventing the international spread of plant pests is the increasing pace of globalization as shown by the international movement of people, goods and pests. For example, around 10 million people pass through Australia's international air ports each year – a number around half the Australian population! In addition around 146 million mail articles, 35,000 vessels and 1.6 million containers arrive in Australia each year. All of these people and items present some risk of pest spread. There is also the challenge of global climate change with trends already suggesting that the risks of pest spread may change significantly. Against this background is the urgent need to overcome the nutritional shortfall of a significant proportion of the world population.

Dramatic examples of new and emerging pests that have spread across international borders to adversely affect agriculture and the environment abound. Recent examples include *Phytophthora ramorum* (sudden oak death), potato spindle tuber viroid and *Candidatus* Liberibacter causing citrus greening. When it comes to pest spread no country is an island – even those like Australia that are islands!

It is sometimes assumed that pest spread is an unavoidable "side effect" of international trade in goods and there is little than can be done to prevent this. However, Australia has found that trade in products where the quarantine conditions are developed and implemented according to the principles and practices of the IPPC are low risk for the introduction of new pests. While it is often difficult to determine the precise pathway of entry of new pests it appears that most new pest incursions in Australia are not related to commercial trade but occur through natural movement from countries to Australia's north, through illegal activity such as travellers bringing in fruit or plant cuttings or occasionally as contaminants of non-agricultural products. Australia's experience is that mainstream trade in agricultural products subject to quarantine risk management measures is a relatively low risk pathway.

While the IPPC operates in a complex international environment interacting with countries and other international agreements it is fundamentally about the practice of plant protection. There are many opportunities for scientific and technical input by plant pathologists into IPPC activities that can make a real difference. Some of these are outlined in the rest of this paper.

Pests do not respect country boundaries so it makes sense to deal with pests on a regional basis. The IPPC recognizes regional plant protection organizations (RPPO) that act as a focus for action on pests and the development of regional standards. While some of them are well resourced and very active others struggle to get enough resources and technical input to make significant progress.

Judging by the number of publications, predicting the invasion of organisms is a popular research topic. Many of these papers take the obvious approach of trying to assess the biological characteristics of successful invaders and use these to predict the success of other organisms. Many of these papers reach the equally obvious conclusion that invasion success is highly dependent on the specific characteristics of the organism and the specific characteristics of the area that is invaded. Very few of the studies yield general principles that can be used to predict how and where and when the thousands of pests that could potentially move internationally will establish.

In many cases the only substantial conclusion is that the best predictor of invasion success is previous invasion success. There is a great deal of scope for lateral thinking in this areas – for research that focuses on the unknown rather than the known, on groups rather than individuals, important general characteristics rather than specific details to provide tools that have practical application for countries trying to manage pest risks associated with trade and the movement of people.

While there have been rapid advances in diagnostic technology over the last 20 years, most of these new techniques are too expensive, too complex, too slow or too specific to be of direct practical use to those managing the risks of pest spread. Fast and inexpensive screening methods are needed that can be used by developing countries with the minimum of training and support. These methods should identify groups of pests rather than individual species and as far as possible be based on biological characteristics that are important for pathogenicity rather than taxonomy.

The impact of pests varies widely depending on factors such as the host range, the environment and the production systems in the area invaded with reported impacts for individual pests often varying by at least an order of magnitude. Methods to estimate the potential impact of a new pest in a new area are important to provide support for planning and determining priorities for action. Research in this area could benefit from a multidisciplinary approach as pest impacts often extend beyond plant health to include trade impacts.

There are many developing countries that lack significant plant health specialists. This lack of specialist expertise often means that these countries have a poor understanding of their plant health status and therefore find it difficult to negotiate quarantine conditions for their exports. There is a great need for training and technology transfer to developing countries in order to improve this situation, an area offering tremendous scope for plant pathologists to make a significant difference.

In summary, the IPPC provides a comprehensive frame work for preventing and controlling pest spread but it is up to plant health professionals to turn this framework into an active and functioning international system that makes a real difference to world food security and the economic progress of developing countries.

Reference

FAO (1999) International Plant Protection Convention. Food and Agriculture Organization of the United Nations

Chapter 12
Pest Risk Analysis as Applied to Plant Pathogens

Françoise Petter, Sarah Brunel, and Muriel Suffert

12.1 International Context of Plant Health

Legally binding rules for international trade have been developed by governments in the World Trade Organization (WTO) and affect trade in plants and plant products. The WTO consists of 153 member countries, which meet regularly in WTO committees to discuss and resolve trade issues, including problems relating to trade in plants and plant products. Two WTO agreements are particularly relevant to this trade: the Agreement on Agriculture and the Agreement on the Application of Sanitary and Phytosanitary Measures (SPS Agreement). In the SPS Agreement it is stated that countries are allowed to restrict trade in plant and plant products when it is necessary to protect plant health. Nevertheless, such restrictions should be based on international standards, guidelines and recommendations, in order to avoid restrictions being used as a barrier to trade. The International Plant Protection Convention (IPPC) is recognized by the WTO agreement as the relevant international standard setting organisation for Plant Health matters. The IPPC has developed several standards, in particular standards on Pest Risk Analysis (https://www.ippc.int/IPP/En/default.jsp). It has also developed the Glossary of phytosanitary terms (ISPM no. 5, IPPC 2007). Terms used in this article are as defined in this glossary.

12.2 EPPO Context and Procedures for Performing PRAs

The European and Mediterranean Plant Protection Organization (EPPO) is an intergovernmental organization responsible for cooperation in plant protection in the European and Mediterranean region. As of October 2008, it has 50 member countries. It is in charge of performing Pest Risk Analyses on a regional scale. While contributing to the development of the International Standard on Phytosanitary

F. Petter(✉), S. Brunel, and M. Suffert
EPPO/OEPP, 1, rue le Nôtre, 75016, Paris, France
e-mail: hq@eppo.fr

R.N. Strange and M.L. Gullino (eds.), *The Role of Plant Pathology in Food Safety and Food Security,* Plant Pathology in 21st Century, 3, DOI 10.1007/978-1-4020-8932-9_12,

Measures no. 11 "*Pest risk analysis for quarantine pests including analysis of environmental risks and living modified organisms*" (ISPM no. 11, IPPC 2007a), EPPO has also developed a regional scheme for PRA now called the *EPPO Decision-support scheme for pest risk analysis of quarantine pests* (EPPO, 2007) (called EPPO Decision-support scheme for PRA hereafter). Compared to ISPM no. 11, this scheme has the added value of guiding the assessor through a logical sequence of questions covering all elements mentioned in this ISPM.

PRAs prepared in the EPPO framework are conducted following the EPPO Decision-support scheme for PRA. The output of a PRA takes the form of a general recommendation to countries, with measures proposed for each organism concerned, distinguishing different levels of risk for different parts of the EPPO region as applicable. This recommendation has then to be adopted by consensus by the EPPO Members, after appropriate consultation. Members decide individually whether the reported risks concern them and select appropriate measures if they do. The EPPO Convention creates no greater obligation on members than that they should "endeavour to implement" EPPO recommendations. However, there is a general policy of "regional solidarity", by which members take phytosanitary measures against pests which are not present in the EPPO region and select their measures from those recommended. Countries may not apply these measures if the risk of establishment on their territory is very low, for example if climatic requirements are not met. The PRA documents reviewed and elaborated in the EPPO framework (PRA records, PRA reports, datasheets) are freely available on the EPPO website (www.eppo.org).

In recent years, a formal process of PRA was established within EPPO as a basis for decision-making on quarantine pest status: the Panel on Phytosanitary Measures reviewed PRAs submitted to it in relation to candidates for the EPPO lists. These PRAs were prepared by individual members. The Panel on Forestry Quarantine Pests itself performed PRAs on its pests of concern. Initially, EPPO had operated mainly on the principle that the organization provided the tools for PRAs to be carried out by NPPOs.

Since 2006, a new organization for the preparation of PRA at the EPPO level.

In 2004 and 2005, the role of EPPO in PRA was discussed both at the political level and at the technical level. It was recognized that many countries do not have the resources to perform PRA and consequently member countries wished that EPPO should play an active role in organizing internationally conducted PRA in the region in order to share costs and workload and to provide technical justification for the regulation of certain pests. The proposal that special EPPO expert groups should now carry out a joint PRA process as well as reviewing PRAs from other sources was made. It was also thought that the creation of a specialized Expert Group would encourage collaboration between members and increase the quality of the PRAs produced. Finally, in September 2005, the EPPO Council formally approved the creation of special Expert Working Groups for PRA that will conduct PRA and terms of references were adopted. Unlike other groups in EPPO, the Expert Working Groups for PRA have a varying composition and experts on specific pests can be called upon to participate when needed. The Expert Working Groups also have core-members to provide consistency. These core-members are usually drawn from existing Panels with experience in performing or reviewing risk assessment and determining risk management options such as the Panel on Phytosanitary Measures and the Panel on Quarantine Pests for Forestry. For each Expert Working Group the EPPO Secretariat tries to balance the experience within the group. Groups include experts on the pest or group of pests to be studied, experts in risk management, experts on the crop concerned, experts with knowledge in running the EPPO PRA scheme, experts on tools to help in assessing the future distribution of the pest (e.g. GIS, Climex) and, whenever possible, an expert in socio-economics, although this has proved difficult to achieve. Usually, experts from the area of origin of the pest are invited so that the group can benefit from their practical experience with the pest.

Experts help the EPPO Secretariat with gathering the necessary information for preparation of the PRA. One expert acts as a steward for the PRA and prepares a pre-PRA so that essential missing information is identified before the meeting of the Expert Working Group takes place.

The Pest Risk Analysis is performed during the meeting of the Expert Working Group for PRA, following the EPPO Decision-support scheme for quarantine pests. The Expert Working Group goes through each individual question on the scheme. Each answer should be justified and justifications are recorded in a document named PRA record which is prepared during the meeting. A datasheet on the pest should also be prepared, preferably by one of the experts on the pest.

The records of the EPPO Decision-support scheme, are sent by email to a group of reviewers who are the core members. When comments are made, the Expert Working Group is consulted by email. The documents are subsequently presented to the relevant bodies which decide by consensus on the appropriate recommendation to be made to EPPO member countries (see previous section). Most examples used in this article are the outcome of these Expert Working Groups.

12.3 Application of the EPPO Decision-Support Scheme on PRA to Pathogens

As outlined above, EPPO has developed a scheme for PRA for quarantine pests. Since PRA is a technical analysis providing a basis for administrative and legislative decisions, it is important that it should be done transparently according to accepted standards. The scheme provides detailed instructions, for the following stages of PRA for quarantine pests: initiation, pest categorization, probability of introduction, assessment of potential economic consequences and pest risk management.

Basically, the PRA is a framework for organizing biological and other scientific and economic information and using it to assess risk. This leads to the identification of management options to reduce the risk to an acceptable level. PRAs can be very short and simple, or very long and complex. There is no fixed criterion for the quantity of information needed. The evaluation does not necessarily have to be quantitative and it can include qualitative considerations, as long as it is scientifically sound (Burgiel et al. 2006). Expert judgement may be used in answering the questions.

The successive stages of the scheme are reviewed in the following part of the article, with examples of how it has been applied to plant pathogens (fungi, bacteria). The scheme follows the sequences presented below.

12.3.1 Initiation

Initiation (see Fig. 12.1) aims to identify the pests or pathways to be considered for risk analysis in relation to the identified PRA area. The EPPO scheme is primarily concerned with the assessment of individual pests, since this is the basis on which European countries formulate their phytosanitary regulations.

ISPM no. 11 also provides for PRA of a pathway. Countries which prohibit the import of most plants and plant products frequently have to consider whether a new trade route can be opened for a previously prohibited plant. The PRA then concerns all the pests which might be carried by this new pathway. A PRA on plants for planting of *Eucalyptus* spp. is on the EPPO work programme for 2009 (following the recent introductions of new pests of Eucalyptus in the EPPO region).

In performing PRAs for individual pests, it is important to establish that their identity is clear. The pests should, as far as possible, be documented before the PRA starts. The information generally needed is listed in EPPO Standard PM 5/1(1) *Check-list of information required for pest-risk analysis* (EPPO 1998). While answering the questions on the scheme, the user should specify all details which appear relevant to the replies to individual questions, indicating the source of the information (Schrader 2005).

12.3.2 Pest Risk Assessment

A Pest Risk Assessment (see Fig. 12.2) consists of determining the probability of different "events" that may lead to the introduction of a pest: what is the likelihood that the pest will enter, establish and spread? The final step of the risk assessment is to determine what potential economic impact the pest is likely to have. Before a full assessment is made a process called Pest categorization is conducted.

12.3.2.1 Pest Categorization

This rapid qualitative assessment is made, with minimal information, to determine whether the organism meets the criteria of the definition of a quarantine

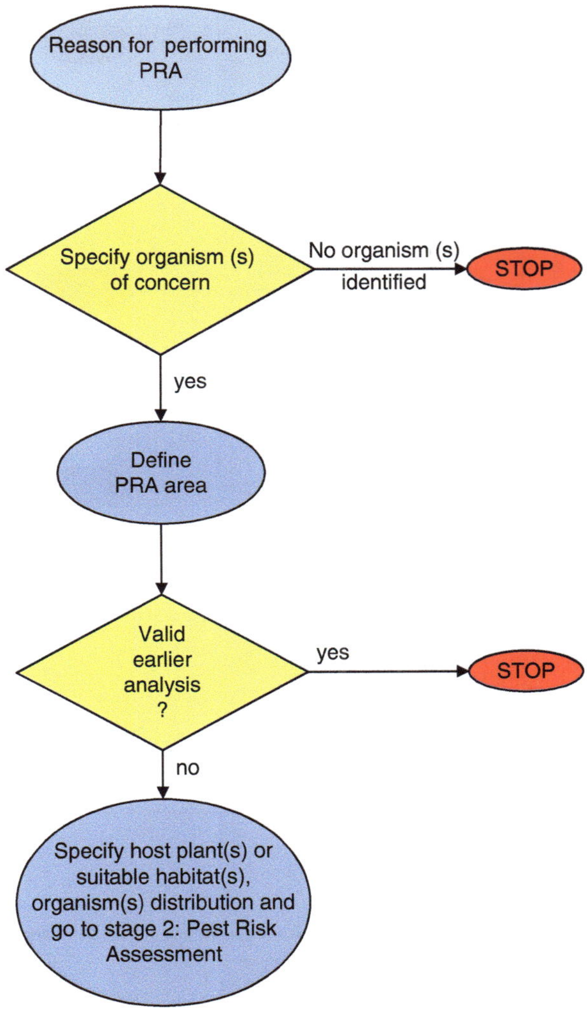

Fig. 12.1 Initiation of the EPPO Decision-support scheme for PRA

pest (*a pest of potential economic importance to the area endangered thereby and not yet present there, or present but not widely distributed and being officially controlled,* IPPC 2007) and therefore can be regulated in international trade. The main aim of this step is to avoid conducting a full PRA in a case which can quickly be seen not to require one. It can also help in identifying what the critical part of the PRA is, that is the pest is of subtropical origin and there are some doubts whether it can establish under European climatic conditions.

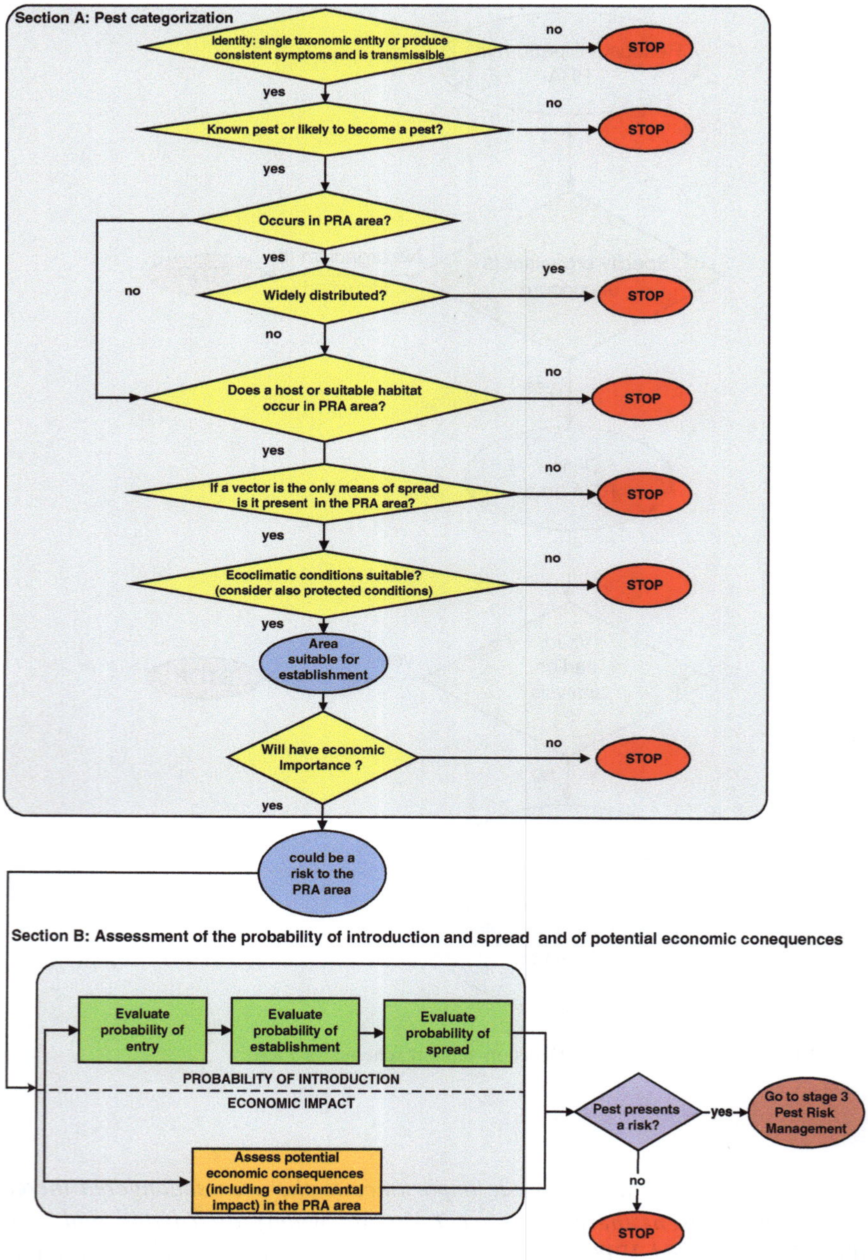

Fig. 12.2 Decision-support scheme for quarantine pests Stage 2 Pest Risk Assessment

However, when the assessor is convinced that the pest clearly presents a risk, this categorization stage can be omitted and the assessor can proceed directly to the main assessment.

If the pest categorization step leads to the conclusion that the pest could present a risk, the main PRA starts. It is essentially composed of a series of questions, to be considered in terms of "likelihood" for qualitative questions (very unlikely, unlikely, moderately likely, likely, very likely), and an estimate for quantitative questions (very few, few, moderate number, many, very many).

12.3.2.2 Full Assessment

Probability of Entry

ISPM no.11 and the EPPO Decision-support scheme for PRA consider the relevant pathways by which the pest has a possibility of being introduced. For example when conducting a PRA on *Gibberella circinata* (Pitch canker) the following pathways were identified: seedlings of *Pinus* spp., seeds of *Pinus* spp. and wood of *Pinus* spp. Plants for planting (including seeds) usually present a higher risk of entry than plant products such as fruits and cut flowers. It is important that the different pathways are studied individually so that only those for which a risk of entry exists are considered at the risk management stage. When a pest has a potential for natural spread (e.g. wind dispersal for rusts) this is also identified at this stage to evaluate the contribution of natural spread to the risk of entry. But as stated in ISPM no. 11, "Measures are not justified if the risk is already acceptable or must be accepted because it is not manageable (as may be the case with natural spread)".

When assessing the risk of entry of a pest on a pathway the following should be considered.

Probability of the Pest Being Associated with the Pathway

Probability of the pest being associated with the pathway at origin: for the pathogen *Xanthomonas axonopodis* pv. *allii* (a bacterium of *Allium* spp.) and the pathway of seeds of *Allium* spp. it was considered that when significant disease development is observed in seed production fields in a tropical environment, the pathogen is likely to be present in the seeds. So the association of the bacterium with this pathway was considered likely for tropical origins.

Concentration of the pest in this pathway: in this question one should consider whether practices, mainly in the country of origin, such as cultural practices (e.g. plant protection products, growing conditions) will decrease the concentration of the pest on the pathway. For example it was considered very likely that the concentration of *X. axonopodis* pv. *allii* is high on seedlings of *Allium* spp. due to cultural practices (high plant density and high humidity). In the case of *Phytophthora lateralis*, applying phytophthora-controlling fungicides could mask the presence of *P. lateralis* on nursery stock and increase the risk of introducing the disease. This should be considered as well.

Volume and frequency of the movement along the pathway: although fundamental, such information is extremely difficult to obtain. Information is usually obtained from FAO STAT (2008) or EUROSTAT (2008), but is usually not specific enough (e.g. global figures for plants for planting).

Probability of Survival During Transport or Storage

For pathogens, the potential of survival during transport or storage is usually very likely to likely as the pest is usually in its host thus protected from external variable conditions. For example *Tilletia indica* is moved in trade in the form of teliospores contaminating seed or grain, or as bunted grain. Generally speaking, teliospores of *T. indica* are robust structures although there is some evidence that a proportion of spores may fragment in transit. If teliospores of *T. indica* are present in sori or as bunted grains they are likely to survive as intact and viable structures in transport (Sansford et al. 2006).

Probability of the Pest Surviving or Remaining Undetected During Existing Pest Management Procedures

In this question the efficacy of existing phytosanitary measures (e.g. inspection, testing or treatments) that may be required as a protection against other (quarantine) pests and applied in the exporting country or the importing country is considered. For *X. axonopodis* pv. *allii* it was considered that in most EPPO countries there are no specific requirements for vegetable seeds. No symptoms are visible on seeds and no curative procedure is available for the moment. It was then concluded that there was a high probability of contaminated seed remaining undetected after phytosanitary inspections carried out at entrance of a territory.

Probability of Transfer to a Suitable Host

It is important to consider the intended use of the commodity when determining the probability of transfer. Some intended uses usually present less risk than others (e.g. consumption, processing). For *T. indica*, the probability of transfer for the pathogen with seed is clear since planting infected or contaminated seed in arable land will allow it to transfer. The pathway for the pathogen to enter via grain will depend upon the location of the port of entry and the route that the consignment takes post-entry as well as its final destination and intended use. Infected or contaminated grain destined for transport through, or for processing in areas where wheat is grown, poses the highest risk, since teliospores can be released during transportation and by handling operations (Sansford et al. 2006).

The distribution of the commodity throughout the PRA area is also studied. In the PRA for *T. indica*, it was considered that grain of *Triticum aestivum* and x *Triticosecale* will be mainly used either for human or animal consumption. Grain of *T. durum* will be used for pasta production and production of specialist foods. Locations of grain storage facilities, flour mills, animal feed production facilities, and processors such as bakeries and pasta producers, etc. will vary by country but all EU countries have these inland facilities. The movement of imported consignments from the port of entry to these facilities allows the commodity to become distributed throughout the PRA area.

Probability of Establishment

An organism which enters does not necessarily establish. To establish it has to find suitable conditions to survive and multiply. The probability of establishment is evaluated by studying the following factors.

Availability of Suitable Hosts, or Vectors in the PRA Area

The first parameter necessary for the establishment of the plant is the presence of suitable hosts. They are listed and their number and distribution are assessed to determine whether the pathogen will find an adequate environment in which to become established. If the pest needs a vector it is important to know if this vector is present in the area or not. For example, *Thrips tabaci* a vector of *Iris yellow spot virus* is common in the EPPO region so the virus can be transmitted to other plants and thus establish (and also spread see next section).

Suitability of the Environment

Similarity of climatic conditions in the PRA area and in the current area of distribution of the species is considered. When possible, a climatic prediction analysis is performed (e.g. with the software CLIMEX). Guidance on performing climatic prediction and on the interpretation of maps has been elaborated in order to avoid misunderstanding and misinterpretation. The guidance warns of the limitations of the software, on preliminary work needed before using CLIMEX and provides a method to rigorously produce a map (Baker et al., 2008). Climatic prediction allows the assignment of different levels of risk, according to climatic parameters. For instance, for *X. axonopodis* pv. *allii*, Mediterranean countries are considered more at risk than temperate countries, and northern countries are not considered at risk.

Other abiotic factors such as soil type may be important for some pathogens and should also be considered.

Prevention of establishment by competition and natural enemies is considered. Recent experience in conducting PRA has shown that experts do not consider that establishment can be prevented by competition and natural enemies and it is debated as to whether this question should be maintained.

Cultural Practices and Control Measures

Two elements are assessed: to what extent is the managed environment in the PRA area favourable for establishment? Could existing control or husbandry measures prevent establishment of the pathogen? For *X. axonopodis* pv. *allii* it was noted that overhead irrigation favours the infection and the spread of the pest. In inoculated onion fields with furrow irrigation the bacterium could be recovered from tail water. In several countries in the EPPO region overhead irrigation is common (it is used as part of the management of thrips). In others furrow irrigation is more important.

Likelihood of Survival Following Eradication Programmes

It is important to consider the epidemiology of the pathogen to answer this question. In the case of *Puccinia hemerocallidis* (a rust attacking *Hemerocallis* spp) eradication was considered difficult once the pathogen is introduced given its epidemiology and its ability to survive as a latent infection. It spreads by means of wind-blown or water-splashed urediospores and the movement of infected plants. However, it was also noted that if detection of infected plants was possible and there were no other hosts in the vicinity, there was a chance that the fungus could be eradicated.

Other Characteristics of the Pathogen Affecting the Probability of Establishment

Biological characteristics such as the reproductive strategy, genetic diversity, adaptability are carefully taken into account.

Probability of Spread

A pathogen which can rapidly spread after establishment presents a much greater risk. An assessment is made of the risk of natural spread including wind or rain dispersal, transport by vectors such as insects or birds, combined with the presence of natural barriers and the quantity of pest to be dispersed. Likeliness of spread by human activities (through movement of plants, soil, irrigation waters, etc.) is also evaluated. The possibility of containing the pathogen is also considered, since some treatments may easily contain a pathogen even if it has established.

Assessment of Potential Economic Consequences

In the case of introduced pests, establishment and spread do not necessarily imply that there is a negative impact. So it is necessary to further evaluate whether there are potential negative economic impacts. Under IPPC definitions, economic impact covers environmental effects as well as social impacts.

Any such effects are documented and evaluated for the current area of distribution of the plant and estimated for the PRA area. This may be done in monetary terms, especially for control costs.

The negative effect of the pest on crop yield and/or quality of cultivated plants or on control costs is considered. In the case of *X. axonopodis* pv. *allii* it was noted that the bacterium can cause significant yield losses and high control costs when conditions are suitable. It negatively affects bulb size because of the destruction of the foliage. In the continental United States, yield losses ranging from 10% to 50% are reported. The negative effect was then considered as major. It should be noted that information on the economic impact of a pest is not always easy to find.

When considering environmental impacts, the following points are evaluated: reduction of keystone species; reduction of species that are major components of ecosystems and of endangered species; significant reduction, displacement or elimination of other species; indirect effects on plant communities (species richness, biodiversity); significant change in ecological processes and the structure and stability of an ecosystem (including further effects on plant species), etc. Such impacts are not always easy to evaluate.

Social impacts of pests may also have to be taken into account. These could include: damaging the livelihood of a proportion of the human population, affecting human use (e.g. water quality, recreational uses, tourism, animal grazing, hunting and fishing). In the case of *P. lateralis* the following elements were considered when evaluating the social impacts: income reduction due to loss of businesses (nurseries had to close down) and loss of wood export markets. In addition, because forests were closed in order to avoid further spread of the pest, tourism decreased.

Degree of Uncertainty and Conclusion of the Pest Risk Assessment

The areas and degree of uncertainty for the different parts of the assessment (entry, establishment, spread and economic impact), should be noted. This ensures transparency of the process and may suggest additional research which could help in completing the PRA or give it more accuracy.

The overall conclusion of the pest risk assessment is to enable a decision to be made as to whether the pest qualifies for quarantine measures. If so, pest risk analysis continues with the selection of risk management options, provided the risk identified is considered unacceptable.

12.3.3 Pest Risk Management

This part of the analysis identifies measures to prevent entry, establishment, or spread of the pest. It explores options that can be implemented: (i) at origin or in the exporting country, (ii) at the point of entry or (iii) within the importing country or invaded area (see Fig. 12.3). The options are structured so that, as far as possible, the least stringent options are considered before the most expensive/disruptive ones, and are consistent with the SPS-Agreement and Plant Health principles (described in ISPM no.1 *Principles of plant quarantine as related to international trade*, IPPC 2007b).

Most commonly recommended measures include those intended to prevent a pathway from being contaminated. These measures are addressed to exporting countries. An example of such measures is the recommendation that citrus orchards should be treated for *Guignardia citricarpa.* Specific growing conditions may also be recommended (for *X. axonopodis* pv. *allii* it was considered that the risk would be reduced if seedlings were produced from seeds free of the bacterium and the plants were grown under protected conditions with no overhead irrigation). The requirement that plants and plant products are grown in a pest-free place of production or pest-free area is also very common. Such measures have been recommended

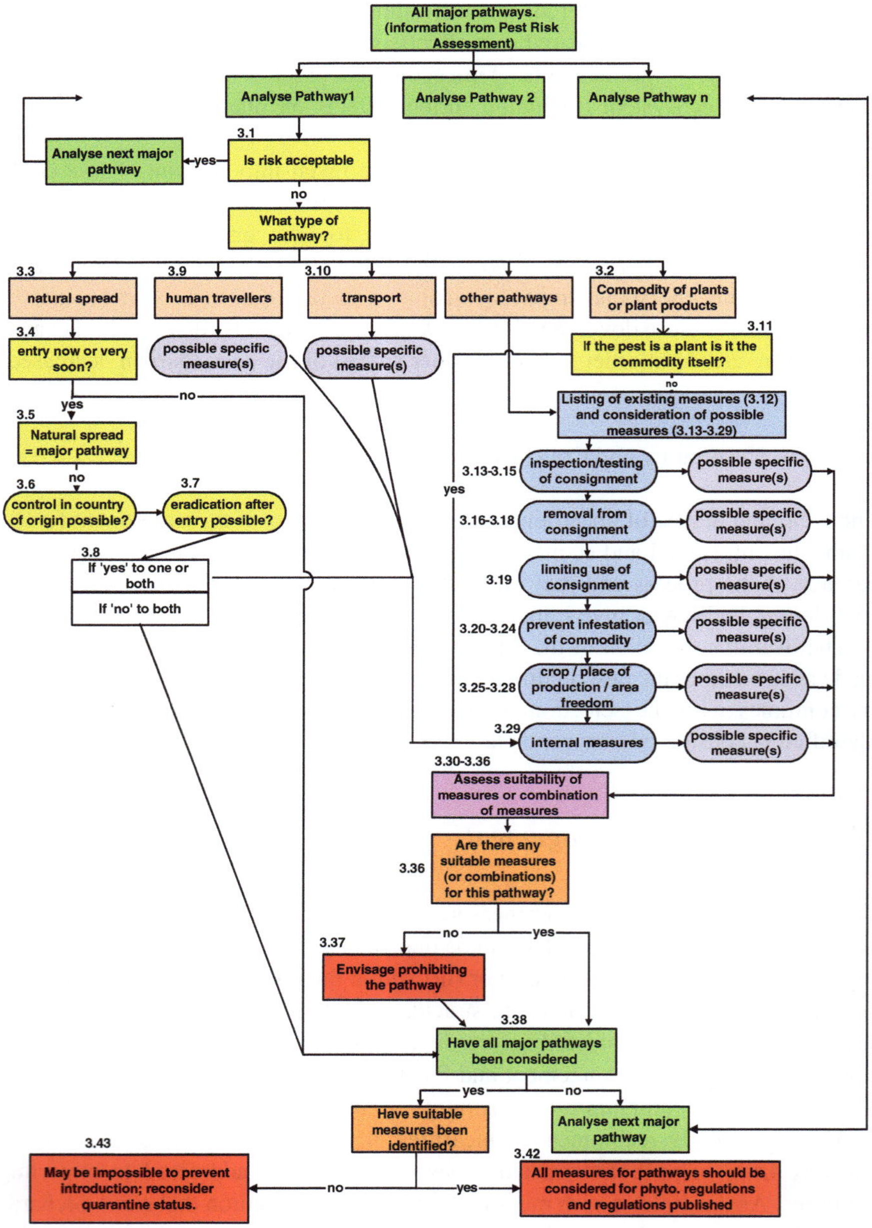

Fig. 12.3 Decision-support scheme for quarantine pests Stage 3 Pest Risk Management

for pathogens that have recently been evaluated such as *P. lateralis*, *X. axonopodis* pv. *allii* and *X. axonopodis* pv. *poinsettiicola* (a bacterium attacking *Euphorbia* sp.). A pest free place of production is a collection of fields, a nursery, greenhouse ....

operated as a single production unit, in which a specific pest does not occur. A pest free area is much larger than a place of production, includes many places of production and may extend to a whole country or parts of several countries.

Measures intended to detect an infestation in a consignment (visual inspection, testing) or treatment of the consignment are also commonly recommended measures.

The scheme also proposes specific measures for entry with human travellers or machinery or means of transport, but they are not very frequent requirements in Europe.

Finally, measures applied when the commodity has entered the country may also be envisaged: such as prevention of establishment by limiting the use of the consignment: import under special licence/permit and specified restrictions. For example the import of seed potatoes from certain provinces of Canada is only permitted in some EU member states, Greece, Spain, Italy, Cyprus, Malta and Portugal, because the risk of establishment of *Clavibacter michiganensis* subsp. *sepedonicus* is considered lower in dry and warm regions (EU 2005).

12.4 Results and Further Challenges

The process of performing PRAs is an effective way to ensure transparency in the adoption of phytosanitary measures. EPPO therefore strongly encourages its member countries to use the EPPO and IPPC Standards on PRA and to train pest risk analysts. In order to encourage collaboration and information sharing, all PRAs, conducted in the EPPO framework, are available on its website (www.eppo.org). EPPO also seeks to constantly improve its scheme and to make it accessible and user-friendly. Therefore a web-based version is under construction. EPPO is also a member of the European Union Framework 7 project PRATIQUE[*] (Enhancements of Pest Risk Analysis Techniques) which aims to develop more efficient risk analysis techniques for pests and pathogens of phytosanitary concern. One of the major challenges of PRATIQUE is the fact that PRAs have very high data demands. Large amounts of information may be required on the pest itself, the situation in its current area of distribution, the pathways of movement, the factors affecting its establishment, spread and impacts in the area under threat and the measures available for its management. Often information on the pest in its area of origin is either lacking, or difficult to access (language, grey literature).

Conducting a PRA requires that the organism to be evaluated is already identified as a pest somewhere (in its area of origin or eventually in its introduced area). This is clearly a limitation as an organism might become a pest only once it is out of its original range. As stated by Brasier (2008) the increasing trade in plants for planting is posing a risk of introduction of organisms for which the potential of being harmful is not known. Some initiatives have taken place and a protocol has been developed in the EU project ALARM[@] to detect new potential pests in their region of origin,

[*] https://secure.csl.gov.uk/pratique/index.cfm

[@] http://www.alarmproject.net/alarm/

before they are introduced into a new area. Sentinel European woody plants have been placed in China and arboreta are being surveyed in Siberia to detect Asian insects and diseases that are particularly damaging to European plant species. This project continues in the framework of PRATIQUE. Such research programmes, which are providing new tools or supporting PRA activities, should be encouraged. The International Plant Protection Convention also recognised the particular risk posed by the increasing trade in plants for planting and an international Standard is being prepared on "Pest risk management for plants for planting in international trade". Finally it is worth underlining that ensuring the phytosanitary security of trade is not solely the responsibility of governments. All actors involved should work together to reach this goal, and this is why other approaches such as Codes of conduct are also being developed.

Acknowledgements We thank Madeleine McMullen for proof reading and Anne-Sophie Roy for her helpful comments.

References

Baker RHA, Black R, Copp GH, Haysom KA, Hulme PE, Thomas MB, Brown A, Brown M, Cannon RJC, Ellis J, Ellis M, Ferris R, Garthwaite R, Glaves P, Gozlan RE, Holt J, Howe L, Knight JD, MacLeod A, Moore NP, Mumford JD, Murphy ST, Parrott D, Sansford CE, Smith GC, St-Hilaire S, Ward NL (2008) Developing a risk assessment scheme for all UK non-native species. Neobiota 7:46–57

Brasier CM (2008) The biosecurity threat to the UK and global environment from international trade in plants. Plant Pathol 57:792–808

EPPO (1998) Pest risk analysis: check-list of information required for pest risk analysis (PRA). EPPO standard PM 5/1(1). www.eppo.org

EPPO (2007) Decision-support scheme for quarantine pests. EPPO Standard PM 5/3 (2). www.eppo.org

IPPC (2007a) Principles of plant quarantine as related to international trade. ISPM no. 1 in International Standards for Phytosanitary Measures, pp 17–26. IPPC Secretariat, FAO, Rome (IT). https://www.ippc.int/IPP/En/default.jsp. Cited 05 Nov 2008

IPPC (2007b) Pest risk analysis for quarantine pests including analysis of environmental risks and living modified organisms. ISPM no. 11 in International Standards for Phytosanitary Measures, pp 135–160. IPPC Secretariat, FAO, Rome (IT). https://www.ippc.int/IPP/En/default.jsp. Cited 05 Nov 2008

Sansford C, Baker R, Brennan J et al., 2006. Pest Risk Analysis for Tilletia indica for the European Union. EU Karnal Bunt Risks Project. Deliverable Report 6·1 and 6·5. http://karnalpublic.pestrisk.net/. Cited 05 Nov 2008

Index

A

Abdalla, M.Y., 100
Agroterrorism, 118–119. *See also* Bioterrorism
ALARM® project, 149–150
Anderson, P.K., 56, 57, 67
Anthony, S., 100
Armengol, J., 100

B

Bacillus thuringienses (*Bt.*)
  Bt cotton, 81–83
  Bt maize, 83–84
Bassi, A., 53
Berkeley, M.J., 53, 59, 60, 62
Biological Weapons Convention, 34
Bio-pesticides
  Bacillus thuringienses (*Bt.*), 81–84
  plant products, 80
  Trichoderma utilization, 81
Biosecurity
  bioterrorism, plant biosecurity
    definitions, 115–116
    evidence, 121–122
    food security linkage, 116
    security through international cooperation, 128–130
    soft targets, 122–123
    technology factor, 123
    threats, 116–121
    vulnerability assessments, 123–128
  definition, 25
  food security and plant diseases, 28
  International regulatory framework
    Biological Weapons Convention, 34
    Cartagena Protocol, 33–34
    Commission on Phytosanitary Measures (CPM), 32
    Convention on Biological Diversity (CBD), 33
    GATT agreement, 32
    International Plant Protection Convention (IPPC), 31–33
    Regional Plant Protection Organizations (RPPOs), 34
    World Organization for Animal Health (OIE), 32
  international travel effects, 29
  issues and challenges
    pest risk analyses and climate change, 38
    phytosanitary measures, 37–38
  National plant health systems
    National phytosanitary measures, 30
    National plant protection organization, 30–31
  plant pests, 26–27
  quarantine, IPPC
    Commission on Phytosanitary Measures (CPM), 133
    Convention on Biological Diversity (CBD), 133
    globalization effect, 134
    International Standards for Phytosanitary Measures (ISPM), 134
    pests, 134–135
    purpose, 133
    regional plant protection organizations (RPPO), 135
  trade related plant biosecurity issues
    maximum residue levels (MRLs), 36
    phytosanitary measures, 35–36
    WTO SPS committee, 34
  world trade effects, 29
Bioterrorism
  evidence
    the Brazilian cocoa case, 121–122
    unsubstantiated cases, 122

Bioterrorism (*cont.*)
- food security linkage, 116
- globalisation threats, 55
- plant biosecurity
  - definitions, 115–116
  - threats, 116–118
- security through international cooperation
  - law and commerce, 130
  - policy, 129–130
  - science, 128–129
- soft targets, 122–123
- technology factor, 123
- threats
  - agroterrorism, 118–119
  - economic damage, 120
  - identification and assessment, 125–127
  - international relations, 120
  - socio-political systems, 120–121
- vulnerability assessments
  - forensic investigation, 127–128
  - mycotoxin-producing fungi, 127
  - strategy, 123–125

Borlaug, N, 43, 44, 49
Brasier, C.M., 149
Brunel, S., 137
Bt cotton, 81–83
Bt maize, 83–84
Buddenhagen, I.W., 56

C

Cartagena Protocol, 33–34
Chakraborty, S., 58
Commission on Phytosanitary Measures (CPM), 32, 133
Conner, R.L., 101
Convention on Biological Diversity (CBD)
- objectives, 33
- quarantine measures, 133

Corpas-Hervias, C., 100
Cozijnsen, A.J., 89

D

Das, S., 75
Desjardins, A.E., 97, 100, 101
Dugan, F.M., 100
Dumroese, R.K., 100

E

Elliott, C.E., 89
Elmer, W.H., 100
Emerging infectious diseases (EIDs)
- case studies, food and cash crops
  - cassava, 61–62
  - cocoa, 63–64
  - coffee, 62–63
  - potato, 59–60
  - wheat, 60–61
- evolution–adaptation, 59
- pathogen pollution, 56–57
- tackling methods
  - plant health, 65–67
  - resistance, 67
- vectors, 58
- weather, 58

Epiphytotics vulnerability, globalisation, 55–56
Epipolythiodioxopiperazine (ETP) toxin. *See* Sirodesmin PL
EU CROPBIOTERROR project, 126
European and Mediterranean Plant Protection Organization (EPPO)
- context and procedures, 137–139
- Decision-support scheme application
  - full assessment, 143–147
  - initiation, 140–141
  - pest categorization, 140–143
  - pest risk management, 147–149

Evans, H.C., 53
Evolution–adaptation, 59

F

Food security in Africa
- bio-pesticides
  - *Bacillus thuringienses* (*Bt* .), 81–84
  - plant products, 80
  - *Trichoderma* utilization, 81
- biotechnology
  - fungal diseases management, 84
  - pests management, 78
  - virus disease management, 84
- disease free planting material, tissue culture, 78–80
- global food situation, 77
- yield losses, diseases and pests, 77–79

Fox, E.M., 89, 94
Fumonisins
- biosynthetic gene (*FUM*) cluster, 106
- effects, 104
- structure, 105

Fusarium proliferatum
- pathogenicity
  - diseases, 99–102
  - genetic substructure, 102–104
  - insecticidal activity, 102
  - phylogenetic tree, 103

symptomology, bakanae disease, 102
secondary metabolism
fumonisins, 104–106
gibberellic acids (GAs), 106
metabolites, 104
taxonomy and identification
DNA sequence comparison, 98–99
morphological characters, 98
sexual mating assay, 98
species and phylogenetic analyses, 97–98
species-specific PCR assays, 99

G
Galván, G.A., 100
GATT agreement, 32
Gene silencing, 75
Genetic modification (GM)
chickpea and black gram, 75
food crisis, 74
pulse network, ISCB, 74
risk crops, 73–74
RNA based plant vaccination, 75–76
Ghiasian, S.A., 100
Gibberellic acids (GAs), *Fusarium proliferatum*
strains, 106
structure, 105
Gliotoxin gene cluster, 92
Global food security
biotechnology application
fungal diseases management, 84
pests management, 78
virus disease management, 84
genetic modification (GM) (*see* Genetic modification)
ISPP (*see* International Society for Plant Pathology)
Globalisation
emerging infectious diseases (EIDs)
case studies, 59–64
evolution–adaptation, 59
pathogen pollution, 56–57
tackling methods, 65–68
vectors, 58
weather, 58
threats
bioterrorism, 55
epiphytotics vulnerability, 55–56
new-encounter pathogens, 54–55
pathogen pollution, 56
Gorbachev, M., 76
Gullino, M.L., 115

H
Hohn, T., 73
Howlett, B.J., 89, 94
Hyun, J.W., 100

I
International Congress of Plant Pathology (ICPP), 43–44
International Plant Protection Convention (IPPC)
convention establishements, 31
number of parties, 33
quarantine measures
Commission on Phytosanitary Measures (CPM), 133
Convention on Biological Diversity (CBD), 133
International Standards for Phytosanitary Measures (ISPM), 134
pests, 134–135
purpose, 133
regional plant protection organizations (RPPO), 135
standard setting programme, 32
International Society for Plant Pathology (ISPP)
challenge programme
Ghana, 46–47
South Africa, 47–48
International Congress of Plant Pathology (ICPP), 43–44
journal, food security, 48–49
plant diseases impact, 48
postgraduate training, 48
poverty and hunger, 44
stark challenge, 50
task force on global food security, 44–46
International Standards for Phytosanitary Measures (ISPM), 134
International travel effects, biosecurity, 29

J
Jiménez, M., 100
Journal, food security, 48–49
Jurjevic, Z., 101

K
Khoury, W., 25
Kingsolver, C.H., 57
Klinkowski, M., 57
Kritzinger, Q., 100
Kuhlman, E.G., 100, 101

L
Láday, M., 101
Large, E.C., 57, 60, 65, 67
Latxague, E., 124, 125
Leslie, J.F., 100, 101
Logrieco, A., 100

M
Madden, L.V., 124
Marín, S., 101
Moretti, A., 97, 100, 101
Moses, E., 11, 46, 47
Mungbean yellow mosaic virus, 75–76
Munkvold, G., 100
Mycotoxins
  Fusarium proliferatum
    pathogenicity, 99–104
    secondary metabolism, 104–106
    taxonomy and identification, 97–99
  sirodesmin PL, epipolythiodioxopiperazine (ETP) toxin
    biosynthesis and regulation, 90–91
    Gliotoxin gene cluster, 92
    origin and distribution, ETP gene, 92
    role, *Leptosphaeria maculans* virulence, 92–94
    structures, 90

N
Nweke, F.I., 11

O
Ocamb, C.M., 101
O'Donnell, K., 99
Onsando, J.M., 77

P
Pathogen pollution, 56–57
Pest risk analysis (PRA), EPPO Decision-support scheme application
  assessment
    degree of uncertainty, 147
    entry probability, 143–144
    establishment probability, 145–146
    pest categorization, 140–143
    spreading probability, 146–147
  challenges, 149–150
  context and procedures, 137–139
  initiation, 140–141
  international context of plant health, 137
  management
    consignment infestation, 149
    pest-free place of production, 147–149
Petter, F., 137
Plant diseases
  crop health syndromes, 6–7
  research to prevent losses
    epidemics and practical implications, 6
    resistance genes identification, 5–6
  rice diseases as yield-reducers
    management tools efficiencies, 5
    responsible diseases, 4
    virus epidemics, 5
  unprecedented agricultural changes and challenges
    challenges, 4
    price spike causes, 3
PRATIQUE project, 149
Proctor, R.H., 97

Q
Quarantine measures, IPPC
  Commission on Phytosanitary Measures (CPM), 133
  Convention on Biological Diversity (CBD), 133
  globalization effect, 134
  International Standards for Phytosanitary Measures (ISPM), 134
  pests, 134–135
  purpose, 133
  regional plant protection organizations (RPPO), 135
Queen Victoria, 60

R
Regional plant protection organizations (RPPOs), 34, 135
Rim, S,-O., 100
Roberts, W., 133

S
Sansford, C.E., 61
Savary, S., 3
Schaad, N.W., 124
Schachermayr, G., 73
Scott, P., 43
Sirodesmin PL
  biosynthesis and regulation, 90–91

Gliotoxin gene cluster, 92
origin and distribution, ETP gene, 92
role, *Leptosphaeria maculans* virulence, 92–94
structures, 90
Small interfering (si) RNAs, 75
Stack, J.P., 115
Stankovic, S., 100
Strange, R., 43, 48, 49
Suffert, F., 115, 124, 127
Suffert, M., 137

T
Tesso, T., 101
Thurston, H.D., 57
Tissue culture, 78–80
Transgene-free gene technology, 75–76
Trichoderma utilization, bio-pesticides, 81
Tsavkelova, E.A., 100

V
Van Alfen, N.K., 57
Van Den Bosch, F., 124
van derGraaff, N.A., 25
Van de Wouw, A.P., 89
Vectors, 58
Veluthambi, K., 75

W
Waalwijk, C., 101, 106
Waller, J.M., 53
Wambugu, F., 77
Wilson, C.L., 57
World Organization for Animal Health (OIE), 32
World trade effects, biosecurity, 29

Z
Zeigler, R.S., 3

MIX
Papier aus verantwortungsvollen Quellen
Paper from responsible sources
FSC® C105338

If you have any concerns about our products,
you can contact us on
ProductSafety@springernature.com

In case Publisher is established outside the EU,
the EU authorized representative is:
Springer Nature Customer Service Center GmbH
Europaplatz 3, 69115 Heidelberg, Germany

Printed by Libri Plureos GmbH
in Hamburg, Germany